Robert Boyle

The Origine of Formes and Qualities

According to the Corpuscular Philosophy

Robert Boyle

The Origine of Formes and Qualities
According to the Corpuscular Philosophy

ISBN/EAN: 9783337194994

Printed in Europe, USA, Canada, Australia, Japan

Cover: Foto ©berggeist007 / pixelio.de

More available books at **www.hansebooks.com**

THE
ORIGINE

OF

FORMES and QUALITIES,

(According to the *Corpuſcular Philoſo-phy*,) Illuſtrated by *Conſiderations* and

EXPERIMENTS,

(Written formerly by way of *Note* and an *Eſſay* about NITRE)

By the Honourable
ROBERT BOYLE,
Fellow of the *Royal Society*.

Audendum eſt , & Veritas inveſtiganda ; quam etiamſi non aſſequamur, omnino tamen propius, quàm nunc ſumus, ad eam perveniemus. Galen.

OXFORD,
Printed by H. HALL Printer to the University,
for RIC: DAVIS. An. Dom. MDCLXVI.

Novemb. 2. 1665.

Imprimatur

ROBERTVS SAY,

Vicecancellarius

O X O N.

The Publisher to the Ingenious Reader.

*I*N *this curious and inquisitive Age, when men,
altogether diffatisfied and wearied out with
the wranglings and idle speculations of the
Schools, are with equal zeal and industry so ear-
nest in their quest and pursuit of a more solid,
rational, and useful* Philosophy *, it may prove a
work very obliging and meritorious to help and
guide them in their studies and researches, and
to hang out a Light to them, (as the Ægyptians
used to do from their highly celebrated* Pharos,
*for direction to the Mariners, that sailed in
those dangerous Seas neer* Alexandria,*) whereby they may, with better success, steer their course
through the vast* Ocean *of Learning, and make
more full and perfect Discoveries of hitherto unknown* Philosophical *verities: which has been
the chief Design of this* Gentleman of Honour,
the most excellent *and* Incomparable *Author
in this* Treatise *now presented to your view,
wherein Principles are not (as was the mode and
guize of former times) obtruded on the World
upon the account of a* Great Name, *or involved
in cloudy and mystical Notions, which put the*
Understanding *upon the Wrack, and yet when*

A 2   *with*

with all this labour and toile of the Brain they are at last known, prove impertinent and use-lesse to the making out with satisfaction, or so much as tolerably, the ordinary Phænomena, *which* Nature *every day presents the world with, but such as are built upon the firme and immoveable foundation of Reason, Sense, and Experience, plain and obvious as well to the Eye as the* Understanding, *and no less accurate and certain in their Application. And though the most* noble Author *hath herein, for the main, espoused the* Atomical Philosophy *(corrected and purged from the wild fancies and extravagancies of the first* Inventours *of it, as to the* O-rigine of the Universe, *and still imbraced with so much kindness and tenderness by some* Pretenders, *against which He hath so Learnedly disputed in his first part* Of the Usefulness of Ex-perimental Philosophy, p. 74. &c.) *in expli-cating the Appearances; yet considering the several Alterations and Additions (the happy pro-duct of his penetrating judgment) made therein, I may not scruple to call it a New* Hypothesis, *peculiar to the* Author, *made out by daily Ob-servations, familiar Proofs and Experiments, and by exact and easily practicable* Chymical *processes, whereby one of the most abstrusest parts of* Natural Philosophy, *the* Origine of

Forms

. The Publisher to the Reader.

Forms and Qualities, *which so much vexed and puzzled the* Antients, *and which, I would speak with the leave of the* Cartesians, *their* Ingenious Master *durst scarce venture upon, or at least was unwilling to handle at large, is now fully cleared, and become manifest: so that from this very* Essay *we may well take hope, and joyfully expect to see the noble* Project *of the famous* VERULAM (*hitherto reckond among the* Desiderata) *receive its full and perfect Accomplishment, I mean, a real, useful, and experimental* Physiology *established and bottomed upon easi,true, and generally received* Principles. *But I shall not forestall thy judgment either about the Excellency of the Author, or his Subject, who hath so freely communicated to the* World *those* treasures *of* Learning, *wherewith his Mind is enriched, but shall soon refer you to the* Work *it self, after I have given you these few Advertisements.*

The following Discourse (*as is easily perceivable by divers Passages thereof) being written, several years since, whole and entire, as now it is, I know not whether it will be worth while to intimate, that the* Author, *casually turning over of late a very recent* Chymical Writer, *found in one of his Treatises (divers of which he never to this day read over) a part of the* Fifth Expe-

a 3
riment

ment *of the second Section; but, as He professes,* (*and sure is like to be believed,*)*he did not dream that That* Chymist, *or any other Author what-soever had lighted on that part of the* Experiment *till a good while after he had made and examined That, among many others, concerning* Salts, *as may be easily guess'd by the peculiar uses and applications He made of it. And though He had met with so unlikely an Experiment in a Writer, who, whether he deserve it or no, has the ill fortune to be much accus'd of Insincerity, and some of whose more easie processes our Author* (*who yet is willing to spare his* Name, *and seems to think his works not useless*) *could not find to succeed, He should not have taken it upon his* Authority, *no more then be is wont to take other* Processes, *divers of which He yet in the general supposes may be true upon the relation of other Chymists; who by blemishing their Books by things untrue and justly suspicious, are not to be relyed on, nor much thanked by wary men. But twill probably appear lesse pertinent to adde any thing further on this subject, then to take notice, that when the* Author *had once consented to the Publication of the following* Papers, *He several times wish'd for an Opportunity to make the* Experiments *and* Observations, *He now presents to the* Publick,

more

The Publisher to the Reader.

more full and compleat, then they were when addres'd to a private Friend. *But the* Contagion, *that drove him from the Places, where his Accommodations for repeating Experiments were, oblig'd Him to apply Himself to other* Studies *and* Employments.

And upon the same account, though he afterwards found many of his Notes *upon other parts of the Essay of* Salt-petre, *and have lying by him divers* Papers *concerning* Sensible Qualities, *and* Sensation in general, *and the* Production *of* Second Qualities, *together with a collection of* Notes about Occult Qualities, *and some other Subjects of kin to those of this* Book; *yet having, upon the freshly intimated Occasion, diverted his Thoughts to other Subjects, He will not engage himself to put together and communicate his Collections on these Subjects by any* Publick *promise.*

Onely thus much perchance I may undertake for, if a fair Opportunity offer it self, that the Author may be induc'd to adde ere long, for the completion of this present Work, *a* Discourse *of* Subordinate Forms, *wherein He, not finding that they have been by any one attempted to be explicated by the* Corpuscularian Hypothesis, *hath proposed an Account of them agreeable thereunto.*

Furthermore, as the Author *has in the follow-*

ing

*ing Diſquiſitions aim'd not at the raiſing or a-
betting a Faction in Philoſophy, but at the Diſ-
covery of the Truth; ſo he is not ſo ſollicitous
what every ſort of Reader will think of his Attempts, (which 'tis eaſie to foreſee are not like to
be overwelcome to the Votaries of the* School
Philoſophy) *as to refuſe a Compliance with the
deſires of his Friends, who have been long ſince
very earneſt with him not to ſpend that time in*
Replies *to particular Perſons, which might be
more uſefully imploy'd in purſuing further Diſcoveries of Nature by* Experiments. *If he meet
with any cogent and material Objections againſt
any of his chief Opinions, He is enough a Lover
of Truth, to be diſpoſ'd to think himſelf oblig'd
by thoſe that ſhall ſhew him his Miſtakes, and to
take occaſion to reforme them. But if nothing*
new or weighty *be urg'd, He conſiders, that he
lives in an Age, wherein he has obſerv'd (even
in his Own caſe) that Truths, if recommended
by real Experiments, will in time make their own
way, and wherein live ſtore of Ingenious Men,
who, for the main, approve the Opinions, and
probably will not diſlike the Arguments he has
propoſ'd, and who being more at leiſure then He
to write Polemical Books, will not ſilently ſuffer
what they judge Truth, to be triumph'd over, or
oppreſſd by thoſe, who, imploying uſually but*
Scho-

The Publisher to the Reader.

Scholaſtical Arguments, may be confuted by Anſwers of the like nature. And therefore He doubts not, but that ſome Learned Favourers of the Corpuſcularian Philoſophy *(of which he hath endeavour'd to make out thoſe parts, wherein they almoſt all agree) will be both able and willing to defend thoſe Diſcoveries by rational Diſputations, that they have not Opportunity to increaſe by* New Experiments.

In the mean while I have no Temptation to doubt in the leaſt, but that this curious and excellent Piece will be entertained and received by all that have any regard to the great concerns of Learning with that guſt, delight, reſpect, and eſtimation which it ſo highly merits.

✿ ✣✣✣✣✣✣ ✣✣✣✣✣✣✣✣✣✣✣✣✣✣✣

The following Treatiſe being printed in the abſence of the Honourable Author, there hath happened (through the miſplacing of the ſeveral Buidles writ apart fairly for the Preſs) a Diſlocation at the 107. page, (as is there alſo intimated) where the firſt Section of the Hiſtorical part is placed, which ſhould not have come in till p. 169. after the diſcourſe of FORMS.

✣✣✣✣ ✣✣✣✣✣✣✣✣ ✣✣✣✣✣✣✣✣✣✣✣✣✣✣

The Proœmial Discourse to the Reader. *

A S tis the part of a Mineralist both to *dif=
cover* new Mines, and to *work* those that
are already discovered, by separating & melting the Oares, to reduce them into perfect
Metals; so I esteem, that it becomes a Naturalist, not onely to *devise* Hypothefes & Experiments, but to examine and *Improve* those that
are already found out. Upon this consideration
(among other Motives) I was invited to make
the following Attempt, whofe productions
coming to be expof'd to other Eyes, then
thofe for which they were first written, twill
be requisite to give the publick fome Account of the Occafion, the Scope, and fome
Circumftances. And this I shall do the more
fully, becaufe the reafons I am to render of
my way of writing in reference to the *Peripatetick* Philofophy, muft contain Intimati-
ons, which perhaps will not be ufelefs to fome
forts of Readers, (efpecially Gentlemen,) and

---

* The following *Preface* being addrefsd onely to *Pyro=
philus,*

by

by being apply'd to moſt of thoſe other parts of my Writings, that relate to the School Philoſophy, may do Them good ſervice, and ſave both my Readers and me ſome trouble of Repetitions.

Having four or five years ago publiſhd a little *Phyſico-Chymical* Tract about the differing parts and redintegration of Nitre, I found as well by other ſignes as by the Early ſollicitations of the Stationer for a new Edition, that I had no cauſe to complain of the Reception that had been given it: But I obſerved too, that the Diſcourſe, conſiſting chiefly of Reflexions, that were occaſionally made upon the *Phænomena* of a ſingle Experiment, was more available to confirme thoſe in the Corpuſcularian Philoſophy, that had already ſomewhat inquir'd into it, then to acquaint thoſe with the principles and notions of it, who were utter Strangers to it; and as to many Readers, was fitter to excite a Curioſity for that Philoſophy, then to give an Introduction thereunto. Upon this Occaſion it came into my mind, that about the time when I writ that Eſſay about Salt-petre, (which was divers years before twas publiſhed) I had alſo ſome thoughts of a Hiſtory of Qualities, and that having in looſe Sheets ſet

down

down divers Obfervations and Experiments
proper for fuch a Defign, I had alfo drawn
up a Difcourfe, which was fo contriv'd, that
though fome parts of it were written in fuch
a manner, as that they may ferve for Expo-
fitory Notes upon fome particular paffages
of the Effay; yet thofe parts with the reft
might ferve for a *General Preface* to the Hiftory of Qualities, in cafe I fhould ever have
Conveniency as well as Inclination to make
the profecuting of It my Bufinefs; and in the
mean time might prefent That *Pyrophilus*,
to whom I writ, fome kind of Introduction to
the principles of the Mechanical *Philofophy*,
by expounding to him, as far as my Thoughts
and Experiments would enable me to do, in
few words, what, according to the Corpufcularian Notions, may be thought of the Nature and Origine of Qualities and Forms; the
knowledge of which either makes or fuppo-
fes the moft fundamental and ufeful part of
Natural *Philofophy*. And to invite me to
make ufe of thefe Confiderations and Tryals
about Qualities and Formes, it opportunely
happen'd, that though I could not find many of the Notes written about particular
Qualities, (my loofe papers having been, during the late Confufions, much fcatter'd by
the

the many Removes I had then occasion to make,) yet when last Winter, being urged to publish my History of Gold, (which soon after came forth,) I rumag'd among my Loose papers, *I* found, that the several Notes of mine that he had met with under various heads, but yet all concerning the Origine of Forms and Qualities, together with the *Pre-*face addrefs'd to *Pyrophilus*, (though written at distant times and places) had two or three years before, by the care of an Industrious person, with whom *I* left them, been fairly copied out together, (which circumstance *I* mention, that the Reader may not wonder to find the following Book not written uniformly in one continued tenor,) excepting some Experiments, which having been of my own making, 'twas not difficult for me to perfect, either out of my Notes and memory, or (where *I* doubted their sufficiency) by repeated Tryals. So that if the Urgency, wherewith divers Ingenious Men prefs'd the publication of my new Experiments about Gold, and my unwillingnefs to protract it, till the Frofty feafon, that was fittest to examine and prove them, were all pafs'd, had not prevail'd with me to let those Obfervations be made publick the last VVinter, they might have

been

been Accompanied with the prefent Effay of the *Origine of Qualities and Formes,* which may be premifd to what *I* have written touching Any of the particular Qualities,fince it containes Experiments and Confiderations fit to be præliminary to them all.

But though *I* was by this meanes diverted from putting out the following Treatife at the fame time with the Hiftory of Gold, yet *I* was without much difficulty prevaild with not to alter my intentions of fuffering it to come abroad; becaufe divers of my Hiftorical Accounts of fome particular Qualities are to be reprinted, which may receive much Light and Confirmation by the things deliver'd in this prefent Treatife about Qualities and Forms in general. To which Inducement was added the *Perfwafion* of fome ingenious *Perfons,*who are pleafed to confeffe their having receiv'd more Information and Satisfaction in thefe *Papers* then *I* durft pretend to give them: though indeed the Subject is fo noble and important, and does fo much want the being illuftrated by fome diftinct and Experimental *Difcourfe,* that *not onely* if *I* did not fufpect my Friends of *Partiality,* I fhould hope that It may gratify many Readers, and inftruct more then a few: but fuch as it is, I

'doe

do not altogether difpair, that it will prove
neither unacceptable, nor ufeleffe. And indeed
the doctrines of Forms and Qualities, and
Generation, and Corruption, and Alteration
are wont to be treated of by Scholaftical
Philofophers, in fo obfcure, fo perplex'd, and
fo unfatisfactory a way, and their Difcourfes
upon thefe Subjects do confift fo much more
of Logical and Metaphyfical Notions and Niceties, then of Phyfical Obfervations and
Reafonings, that it is very difficult for any
Reader of but an ordinary Capacity, to underftand what they mean, and no leffe difficult for any intelligent and unprejudic'd Reader to acquiefce in what they teach: which is
oftentimes fo precarious, and fo contradictious to its felf, that moft Readers (without
alwaies excepting fuch as are Learned and
Ingenious) frighted by the darknefs and difficulties wherewith thefe Subjects have been
furrounded, do not fo much as look after or
read over thefe general and controverted
matters, about which the Schools make fo
much noife; but defpairing to find any fatif-
faction in the ftudy of them, betake Themfelves immediately to that part of Phyficks
that treats of particular Bodies: fo that to
Thefe it will not be unacceptable to have any

Intel-

intelligible Notions offer'd them of thofe
Things, which, as they are wont to be propof'd, are not wont to be underftood:
though yet the Subjects themfelves, if I miftake not, may be juftly reckon'd not onely
amongft the nobleft and moft important, but
(in cafe they be duely propof'd,) among the
ufefulleft and moft delightful Speculations,
that belong to Phyficks.

I confider too, that among thofe that are
inclin'd to that Philofophy, which, I find, I
have been much imitated in calling *Corpufcularian*, there are many Ingenious Perfons,
efpecially among the Nobility and Gentry,
who, having been firft drawn to like this new
way of Philofophy, by the fight of fome Experiments, which for their Novelty or Pret-
tinefs they were much pleaf'd with, or for
their Strangenefs they admir'd, have afterwards delighted Themfelves to make, or fee
variety of Experiments, without having e-
ver had the Opportunity to be inftructed in
the Rudiments or fundamental Notions of
that Philofophy, whofe pleafing or amazing
Productions have enamour'd them of It. And
as Our *Pyrophilus*, for whom thefe Notes
were drawn up, did in fome regards belong
to this fort of *Virtuofi*, fo tis not impoffible,

but

but that such Readers, as He was then, will not be sorry to meet with a Treatise, wherein though my chief and proper businesse be the giving some Account of the Nature and Origine of Forms and Qualities; yet by reason of the connexion and dependance betwixt these and divers of the other principal Things, that belong to the general part of Physicks, I have been oblig'd to touch upon so many other important Points, that this Tract may, in some sort, exhibit a Scheme of, or serve for an *Introduction* into the Elements of the Corpuscularian Philosophy.

And as those Readers, that have had the Curiosity to peruse what is commonly taught in the Schools about Forms, and Generation, and Corruption, and those other things we have been mentioning, and have (as is usual among ingenious Readers) quitted the study of those unsatisfactory intricacies with Disgust, will not be displeas'd to find in our Notes such Explications of those things as render them at least intelligible: so it will not perhaps prove unacceptable to such Readers, to find those matters, which the Schools had interwoven with *Aristotle's* Doctrine, reconcil'd and accommodated to the Notions of the Corpuscular Physicks.

If

If it be faid, that I have left divers things unmention'd, which are wont to be largely treated of by the *Ariftotelians*, and particularly have omitted the Difcuffion of feveral Queftions, about which they are wont very folemnly and eagerly to contend, I readily acknowledge it to be true: But I anfwer further, That to do otherwife then I have done, were not agreeable to the nature of my Defign, as is declar'd in the Preface to *Pyrophilus*; and that though moft Readers will not take notice of it, yet fuch as are converfant in that fort of Authors, will, I prefume, eafily find, that I have not left them unconfulted, but have had the Curiofity to refort to feveral both of the more, and of the leffe recent Scholaftical Writers about Phyficks, and to fome of the beft Metaphyficians to boot, that I might the better inform my felf, both what their Opinions are, and upon what arguments they are grounded. But as I found thofe Inquiries far more troublefome then ufeful, fo I doubt not, that my omiffions will not much difpleafe that fort of Readers, for whofe fake chiefly tis that thefe Papers are permitted to be made publick. For if I fhould increafe the Obfcurity of the Things themfelves I treat of, by adding the feveral Obfcu-

rer

rer Comments (rather then Explications,)
and the perplex'd and contradictious Opinions I have met with among Scholaſtick Writers, I doubt that ſuch perſons, as I chiefly
write for, would inſtead of better compre-
hending what I ſhould ſo deliver, abſolutely
forbear to read it. And there being many
Doctrines, to which number This we are ſpeaking of ſeems to belong, wherein the ſame
innate Light, or other Arguments, that diſcover the Truth, do likewiſe ſufficiently ſhew
the Erroneouſneſs of diſſenting Opinions, *I*
hope it may ſuffice to propoſe and eſtabliſh
the Notions that are to be imbrac'd, without
ſollicitouſly diſproving what cannot be true,
if thoſe be ſo. And indeed there are many
Opinions and Arguments of good repute in
the Schools, which do ſo entirely rely upon the
Authority of *Ariſtotle*, or ſome of his more celebrated Followers, that where that Authority is not acknowledg d, to fall upon a ſo-
lemn Confutation of what has been ſo pre-
cariouſly advanc'd, were not onely unneceſ-
ſary, but indiſcreet even in a Diſcourſe not
confin'd to the brevity challeng'd by the nature of this of Ours. And there are very many Queſtions and Controverſies, which
though hotly and clamorouſly contended a-
                                                    bout,

about, and indeed pertinent and fit enough
to be debated in their Philofophy, do yet fo
much fuppofe the Truth of feveral of their
Tenents, which the new Philofophers reject,
or are grounded upon Technical Terms or
forms of fpeaking, that fuppofe the Truth of
fuch Opinions, or are Expreffions, whereof
we neither do, nor need make any ufe; that
to have inferted fuch Debates into fuch a
Difcourfe as mine, would have been not onely tedious, but impertinent. As (for inftance)
thofe grand Difputes, whether the four Elements are endow'd with diftinct Subftantial
Forms, or have onely their proper Qualities
inftead of them? and whether they remain
in mix'd Bodies according to their Forms, or
according to their Qualities? and whether
the former or the latter of thofe be, or be not
refracted? Thefe, I fay, and divers other controverfies about the four Elements, and their
manner of Miftion, are quite out of doors in
their Philofophy, that acknowledge neithe,
that there are four Elements, nor that Cold,
Heat, Dryneffe, and Moiftore, are in the Peripatetick fenfe, firft Qualities, or that there
are any fuch Things as Subftantial Forms *in*
rerum naturâ. And it made me the more unwilling to ftuff thefe Papers with any needlefs

School.

School-controverfies, becaufe I found, upon
perufal of feveral Scholaftick Writers, (efpecially the recenter, who may probably be
fuppof'd to be the moft refin'd,) that they
do not alwaies mean the fame Things by the
fame Terms, but fome imploy them in one
fenfe, others in another, and fometimes the
fame Writer ufes them in very differing fenfes; which I am oblig'd to take notice of, that
fuch Readers,as have confulted fome of thofe
Authors, may not accufe me of miftaking or
injuring fome of the Scholaftical Terms and
Notions he may meet with in thefe Papers,
when I have enely imploy'd Them in the
fenfe of other School-writers, which I judg'd
preferrible. And this puts me in mind of intimating, That whereas, on the contrary, I
fometimes imploy variety of Terms and Phra-
fes to exprefs the fame Thing, I did it purpofely, though perhaps to the prejudice of
my own Reputation, for the Advantage of
Pyrophilus; both I and others having obferved, that the fame unobvious Notions being
feveral wayes exprefs'd, fome Readers even
among the Ingenioufer fort of them,will take
it up much better in one of thofe Expreffions, and fome in another.

But perhaps it will be wondred at, even by
fome

some of the new Philofophers, That diffen-
ting fo much as I do from *Ariftotle* and the
Schoolmen, I fhould overlook or decline
fome Arguments, which fome very ingenious
Men think to be of great force againft the
Doctrine I oppofe. But divers of thefe Arguments being fuch, as the Logicans call *ad hominem*, I thought I might well enough fpare
them. For I have obferv'd *Ariftotle* in his
Phyficks to write very often in fo dark and
ambiguous a way, that tis far more difficult,
then one would think, to be fure what his
Opinion was: and the Unlearned, and too
frequently jarring Gloffes of his Interpreters,
have often made the Comment darker then
the Text: fo that (though in moft it Be,yet)
in divers cafes tis not eafie (efpecially wuh-
out the expence of many words) to lay open
the Contradictions of the Peripatetick Doctrine; befides, that the urging fuch contradictions are oftentimes fitter to filence an un-
wary Adverfary, then fatisfie a wary and
judicious Reader. It being very poffible,that
a man may contradict himfelf in two feveral
places of his Works, and yet not be in both
of Them in the wrong. For one of his Affertions, though inconfiftent with the other,
may yet be confiftent with Truth. But this

b 4   is

is not all I have to fay on this Occafion. For befides, that having, for many reafons, elfewhere mention'd, purpofely forborn the reading of fome very much, and, for ought I know, very juftly Efteem'd Difcourfes about general *Hypothefes*, tis very poffible, that *I* may be a ftranger to fome of thofe Arguments: befides this, I fay, *I* confefs I have-purpofely forborn to make ufe of others, which I have fufficiently taken notice of. For fome of thofe Ratiocinations would engage him that fhould imploy them, to adopt an *Hypothefis* or Theory, in which perhaps I am not fo throughly fatisfied and of which *I* do not conceive my felf to have, on this occafion, any neceffity to make ufe: and accordingly *I* have forborn to imploy Arguments, that are either grounded on, *or* fuppofe Indivifible Corpufcles, call'd *Atoms*, or any *innate motion* belonging to them; *or* that the Effence of Bodies confifts in Extenfion, *or* that a Vacuum is impoffible; *or* that there are fuch *Globuli caleftes*, *or* fuch a *Materia fubtilis*, as the Cartefians imploy to explicate moft of the *Phænomena* of Nature. For thefe, and divers other Notions, I (who here write rather for the Corpufcularians in general, then any party of them) thought it improper needlefly

to

to take in, difcourfing, *either* againft thofe, to
whom thefe things appear as difputable, as
the Peripatetick Tenents feem to me; *or* for
to fatisfie an ingenious perfon, whom it were
not fair to impofe upon with *Notions*, that I
did not my felf think proper.

. And on the like Account I forbore fuch
Arguments as thofe, that fuppofe, in Nature
and Bodies inanimate, Defigns, and Paffions
proper to Living, and perhaps peculiar to Intelligent Beings; and (fuch as) fome Proofs
that are drawn from the Theology of the
Schools: (which I wifh leffe interwoven with
Ariftotle's Philofophy.) For though there
be fome things, which feem to be of this fort,
(as Arguments drawn from final caufes in
divers particulars that concern Animals,)
which, in a found fenfe, I not onely Admit, but
Maintain: yet fince, as they are wont to be
propof'd, they are liable enough to be queftion'd, I thought it expedient for my prefent
defign to prætermit them, as things that I do
not abfolutely need; though the imploying
fome of Them would facilitate my Task. And
this I did the rather, becaufe I alfo forbear to
anfwer Arguments, that however vehemently and fubtly urg'd by many of the modern
Schoolmen of the Roman Catholick Commu-
nion

nion, are either confeffedly, or at leaft really
built upon fome Theological Tenents of
theirs, which, being oppof'd by the Divines of
Other Churches, and not left unqueftion d by
fome Acute ones of their Own, would not be
proper to be folemnly taken notice of by Me,
whofe Bufinefs, in this Tract, is to difcourfe of
Natural Things as a Naturalift, without in-
vading the Province of Divines, by inter-
medling with Supernatural Myfteries, fuch as
thofe, upon which divers of the Phyfico-Theo-
logical Tenents of the Schoolmen, efpecially
about real Qualities, and the Separablenefs
of Accidents from Subjects of Inhæfion, are
manifeftly, if not alfo *avowedly*, grounded. But
to return to the other things I was owning to
have left unmention'd; notwithftanding all
that I have been faying, I readily acknow-

ledge,

ledge, that in fome recent Authors, that have been imbracers of the new Philofophy, I have met with fome paffages, that might well and pertinently be taken into the following Dif-courfe, but that having been (as I formerly intimated) tranfcrib'd fome years ago, I can-not now fo conveniently Alter it: which I am the leffe troubled at, becaufe thefe few addi-tional Arguments, thought fit to illuftrate or confirme, being not neceffary to make out what has been deliver'd, may fafely be let a-lone, unleffe there happen (as tis not unlike-ly there may) an occafion of reprinting thefe Notes, with fuch Enlargements as may make them the more fit to be an Introdunction into the Corpufcular Philofophy.

I hope then upon the whole matter, that I have pitch'd upon that way, that was the moft conducive to my Defign, *partly* by infifting onely on thofe Opinions, whether true or falfe, which, for their Importance or Difficul-ty, feem'd to deferve to be particularly either explicated or difprov'd; and *partly*, by choo-fing to imploy fuch Arguments as I thought the cleareft, and cogenteft, and by their affu-ming the leaft of any, feem'd the eafieft to be vindicated from Exceptions: without trou-bling my felf to anfwer Objections, that ap-

pear'd

peat'd rather to be drawn from Metaphyfical or Logical Subtleties, or to be grounded upon the Authority of men, then to be Phyfical Ratiocinations, founded upon Experience, or the nature of the Things under debate; efpecially having, in the propofal and confirmation of the Truth, fo laid the grounds, and intimated the wayes of anfwering what is like to be colourably objected againft it, that an Ingenious man may well enough furnifh himfelf with Weapons to defend the Truth, out of the Notions, Hints, and Experiments, wherewith in this Tract care has been taken to accompany it. And my forbearing to profecute fome of the Peripatetick Controverfies any further then I have done, will not, I hope, be blam'd by Them, that have obferv'd as well as I, how much thofe Difputes are wont to be lengthned by fuch frivolous Diftinctions, as do not deferve to be folemnly examin'd, efpecially in fuch a Compendious Treatife as Ours. For an attentive Reader needs not be much converfant with the writings of the modern Peripateticks, about fuch fubjects as Subftantial Forms, Generation, Corruption, &c. to take notice, that tis their Cuftome, when they find Themfelves diftrefs'd by a folid Argument,

to

to endeavour to elude it by some pittiufll Distinction or other, which is usually so ground-
lesse, and so unintelligible, or so nugatory, or
so impertinent to the Subject, or at least so
insufficient for the purpose tis alleadg'd for,
that to vouchsafe it a sollicitous Confutation
might question a Writers Judgment with intelligent Readers; who by such insignificant
Distinctions are satisfy'd of nothing so much,
as that the Framers of them had rather say
(that which Indeed amounts to) nothing,
then not seem to say something. And of such
Evasions they may probably be emboldned
to make use, by the practice of *Aristotle* himself, to whom such obscure and unsatisfactory
Distinctions are so familiar, that I remember
one of his own Commentators * (and he one
of the most judicious) could not forbear, upon
a certain Text of his Masters, to complain of

* The Author here meant is the Inquisitive Peripatetick
Cabæus, who in one place hath these words. *ut hanc*
quæstionem solvat, recurrit ad illam distinctionem sibi val-
de familiarem; quâ utitur Aristoteles in tota sua Philoso-
phia, quoties obviam habet aliquam gravem difficultatem,
distinguit enim actu vel potentia, &c. In another these:
―*Quæ est distinctio quædam familiaris Aristoteli, quam*
applicat omnibus rebus, ubi difficultates urgent, & videtur
istis vocibus quasi fatali gladio omnes rescindere difficul-
tatis nodos; vix enim est difficultas, cui non putat se satis-
facere distinguendo actu & potentia.

it.

it, and particularly to take notice, that That one Diftinction of *actu & potentia* runs through almoft all *Ariftotle's* Philofophy, and is imploy'd to fhift off thofe Difficulties he could not clearly Explicate.

By which neverthelefs I would not be underftood to cenfure or decry the whole Peripatetick Philofophy, much leffe to defpife *Ariftotle* himfelf; whofe own Writings give me fometimes caufe a little to wonder, to find fome Abfurdities fo confidently father'd upon him by his Scholaftick Interpreters. For I look upon *Ariftotle* as *one* (though but as *one amongft many*) of thofe fam'd Antients, whofe Learning about *Alexanders* time enobled *Greece*; and I readily allow him moft of the prayfes due to great Wits, excepting thofe which belong to clear-headed Naturalifts. And I here declare, once for all, that where in the following Tract, or any other of my writings, I do *indefinitely* depreciate *Ariftotle's* Doctrine, I would be underftood to fpeak of *his Phyficks,* or rather of the Speculative part of them, (for his Hiftorical Writings concerning Animals I much efteem,) nor do I fay, that even Thefe may not have their Ufe among Scholers, and even in Univerfities, if they be retain'd and ftudied with due cautions

ons and Limitations; (of which I have elfewhere fpoken.)

But to refume the Difcourfe, whence the
Peripatetick Diftinctions tempted me to di-
grefs; by any thing I formerly faid, I would
not in the leaft difparage thofe excellent and
efpecially thofe modern Authors, that have
proteffedly oppofed the *Ariftotelian* Phyficks:
(fuch as *Lucretius, Verulam, Baffo, Des Cartes*
and his Followers, *Gaffendus,* the two *Boots,*
Magnenus, Pemble, Helmont,)nor be thought
to have made no ufe of any of their Cogitati-
ons or Arguments. For though fome of their
Books I could not procure, when I had occafion to have recourfe to Them; and though
the weaknefs of my Eyes difcourag'd me from
perufing thofe parts of others, that concern'd
not the Subject I was treating of, yet I hope
I have been benefitted by thofe i have confulted, and might have been more fo, by the
Learned*Gaffendus's* Little,but Ingenious, *Syn-*
tagma Philofophiæ Epicuri, if I had more fea-
fonably been acquainted with it.

But *whether* we have treated of the Nature
and Origine of Forms and Qualities in a more
comprehenfive way then others, *whether* we
have by new and fit Similitudes, and Examples, and other means rendred it more intel-
telligible

ligible then they have done, *whether* we have
added any confiderable number of Notions
and Arguments, towards the compleating
and confirming of the propof'd *Hypothefis*,
whether we have with reafon difmifsd Arguments unfit to be relyd on, and *whither* we
have propof'd fome Notions and Arguments
fo warily, as to keep Them from being lyable
to Exceptions or Evafions, whereto they were
obnoxious as others have propol'd them,
whether (I fay) we have done all or any of
thefe in the firft or Speculative part of this .
Treatife, we willingly leave the Reader to
judge: But in the fecond or Hiftorical part of
It, perhaps he will be invited to grant, that
we have done that part of Phyficks, we have
been treating of, fome little fervice: fince by
the Lovers of real Learning, it was very
much wifh'd, that the Doctrines of the new
Philofophy (as tis call'd) were back'd by particular Experiments; the want of which I have
endeavour'd to fupply, by annexing fome,
whofe Nature and Novelty I am made believe will render them as well Acceptable as
Inftructive. For though, that I might not an-
ticipate what belongs to other papers, I did
not make the Laft *Section* confift of above a
Decad of them; and though, for the reafons

inti-

intimated in the Advertifements premif'd to them, I did not expreſly mention to *Pyrophi-lus* all that I could have told him about them; yet I have been carefull ſo to chooſe them, and to interweave Hints in delivering them, that a ſagacious Reader, who ſhall have the Curioſity to try them heedfully, and make Reflexions on the ſeveral *Phænomena*, that in likelihood will occurr to him, will (if I miſtake not) receive no contemptible information, as of ſome other things; ſo particularly about the nature of Mixtions, (which I take to be one of the moſt important and uſeful, though neglected and ill underſtood, Doctrines of the Practical part of Phyſicks) and may probably light upon more then he Expects, or I have fully Delivered, and perhaps too more then I Foreſaw.

And though ſome *Virtuoſi*, more conver-ſant perhaps with Things then Books, preſu-ming the Decay of the Peripatetick Philoſo-phy to be every where as great, as tis among Them in *England*, may think that a Do-ctrine, which they look on as Expiring, need not have been ſo ſollicitouſly confuted; yet thoſe that know, how deep rooting this Phi-loſophy has taken (both elſewhere, and parti-cularly) in thoſe Academies, where it has

A   flou-

flourifh'd for many Ages, and in fome of which tis, exclufively to the Mechanical *Phi-lofophy*, water'd and fenced by their Statutes or their Superiours: and he that alfo knows, how much more eafie fome (more fubtle, then Candid) Wits, find it plaufibly to defend an Error, then ingenioufly to confefs it; will not wonder, that I fhould think, that a Doctrine fo advantag'd, though it be too erroneous to be Fear'd, is yet too confiderable to be Defpif'd. And not to queftion, whether feveral of thofe, that moft contemn the favourers of the *Peripatetick Hypothefis*, as the later Difcoveries have reduc'd them to Reform it, be not the leaft provided to anfwer their Arguments: (not to queftion this, I fay,) there are divers of our Adverfaries (mifled onely by Education, and morally harmlefs prejudices) who do fo much Deferve a better Caufe, then that which Needs all their fubtlety without being VVorthy of it, that I fhall think more paines, then I have taken, very ufefully beftow'd, if my Arguments and Experiments prove fo happy, as to undeceive Perfons, whofe parts, too unluckily confin'd to Narrow and Fruitlefs Notions, would render them illuftrious Champions for the Truths they are able fo Subtlely to oppofe; and who might

might questionlefs perform Confiderable things, if they imploy'd as much Dexterity to Expound the Myfteries of Nature, as the Riddles of the School-men; and laid out their VVit and Induftry to furmount the Obfcu-rity of Her works in ftead of that of *Arifto-tles.*

There might be a few other particulars fit to be taken notice of in this *Preface,* but fin-ding that I had already mention'd them in that, which I had addrefs'd to *Pyrophilus,* my Haft makes me willing rather to refer the Reader thither for them, then Alter that, or Lengthen this; (which I fhould think much too Long already, if it were not poffible that it may hereafter prove præliminary to more papers then thefe tis now premif'd to.) So that there remaines but one Advertifement neceffary to be given here, namely, that whereas in the following Notes I feveral times fpeak of the Author of the Effay of Salt-petre, as of a third perfon, the Occafion of that was, That when thefe Notes, and fome about particular Qualities, were written, I had a Defign to make two diftinct forts of An-notations upon that Effay; in the *former* whereof (which now comes forth) I affumed the perfon of a Corpufcularian, and difcourft

at

at that rate: But I had thoughts too (in case God were pleas'd to grant me Life and Opportunity,) to take a *second* Review both of the Treatise it self, and of the Notes on it, and on that occasion to Adde what my riper Thoughts and further Experience might suggest unto me. And that in my Animadversions I might, with the more Freedome and Conveniency, Adde, Explain, Alter, and even Retract, as I should see cause, I thought it not amiss to write them, as if they were made on the VVork of another. By which Intimation the Reader may be assisted to ghess how much I intended in the following Discourse, (in which, as in the *Prefaces* belonging to it, I play the Corpuscularian,) to reserve my self the Freedom of Questioning, and Correcting, upon the design'd Review, any thing deliver'd in these Notes ; and how much more it was in them my design to bring *Pyrophilus* Experiments and Queries to Illustrate obscure matters, then, by hasty Assertions, to Dogmatize about them.

THE

The Præface.

THe *Origine* (Pyrophilus) *and Nature of the Qualities of Bodies, is a Subject, that I have long lookt upon, as one of the moſt Important and Uſefull that the Naturaliſt can pitch upon for his Contemplation. For the Knowledge we have of the Bodies without Us, being for the Moſt part fetched from the Informations the Mind receives by the Senſes, we ſcarce know any thing elſe in Bodies, upon whoſe account they can worke upon our Senſes ſave their Qualities: For as to the Subſtantial Formes, which ſome Imagine to be in all Naturall Bodies, it is not halfe ſo Evident, that there are ſuch, as it is, that the wiſeſt of thoſe that do admit them, Confeſſe, that they do not well Know them.* And as tis by their Qualities, that Bodies act Immediately upon our*

---

*Nego tibi ullam eſſe formam nobis notam plenè & planè: noſtrámque ſcientiam eſſe umbram in ſole. Scaliger : (of whoſe confeſſion to the ſame puipoſe, more are cited hereafter.)

B   *Senſes,*

Senses, so 'tis by vertue of those Attributes likewise, that they act upon Other bodies, & by that action produce in Them, & oftentimes in Themselves those Changes, that sometimes we call Alterations, and sometimes Generation, or Corruption.

And 'tis chiefly by the Knowledge, such as it is, that Experience, (not Art) hath taught Us, of these differing Qualities of Bodies, that we are enabled, by a due application of Agents to Patients, to exercise the little Empire, that we have either Acquir'd or Regain'd over the Creatures. But I think not the contemplation of Qualities more Noble & Useful, then I find it Difficult; For what is wont to be taught us of Qualities in the Schools, is so Slight and ill grounded, that it may be doubted, whether they have not rather Obscured, then Illustrated the things they should have explain'd. And I was quickly discouraged from expecting to learne much from them, of the Nature of divers Particular Qualities, when I found, that except some few, which they tell You in

general

The Preface.

General may be deduced, (by wayes they leave those to guesse at that can,) from those foure Qualities, they are pleas'd to call the First; they confesse, that the rest spring from those Forms of Bodies, whose particular Natures, the judiciousest of them acknowledge, they cannot comprehend. And Aristotle himself not only doth (as we shall see anon) give us of Qualitie in Generall, (which yet seems far more easily defineable, then many a Particular Quality,) no other then such a definition, as is as Obscure, as the thing to be declared by it; but I Observe not without some wonder, that in his eight Books of Physicks, where he professedly treats of the Generall Affections of Naturall things, he leaves out the Doctrine of Qualities; as after him Magirus, and divers other Writers of the Peripatetick Physiologie have done: which (by the way) I cannot but look upon as an Omission, since Qualities doe as well seem to belong to Naturall Bodies Generally consider'd, as Place, Time, Motion, and those other things, which upon that

B 2   *account*

account are wont to be Treated of in the Generall part of Natural Philosophy. The most Ingenious Des Cartes has something concerning some Qualities; but though for Reasons elsewhere express'd, I have purposely Forborn to peruse his Systeme of Philosophy; yet I find by Turning over the Leaves that he has Left most of the other Qualities Untreated of, & of Those, that are more properly call'd Sensible, he Speaks but very Briefly & Generally; rather considering what they do upon the Organs of Sense, then what Changes happen in the Objects themselves, to make them Cause in us a Perception sometimes of one Quality, and sometimes of Another. Besides, that his Explications, do many of them so depend upon His peculiar Notions, (of a Materia Subtilis, Globuli Secundi Elementi, and the like) and These, as it became so Great a Person, he has so Interwoven with the rest of his Hypothesis, that They can seldome be made Use of, without Adopting his whole Philosophy. Epicurus indeed, and his Scholiast Lucretius

tius, *have Given some good Hints concer-*
ning the Nature of some few Qualities.
But beside, that even these Explications
are divers of them either Doubtfull or Im-
perfect, or both, there are many other Qua-
lities, which are left for Others to Treat of.
And this is the Second and Maine Difficul-
ty, which I find in investigating the Na-
ture of Qualities, Namely, that Whatever
be to be thought of the Generall Theoryes
of Aristotle, *or other Philosophers, concer-*
ning Qualities; we evidently Want That,
upon which a Theory, to be Solid and Usefull,
must be Built; I mean an Experimentall
History of them. And this we so Want,
that except perhaps what Mathematicians
have done concerning Sounds, and the
Observations (rather then Experiments)
that our Illustrious Verulam *hath (in some*
few Pages) say'd of Heat, in his short Essay,
De Formâ Calidi; *I know not Any one*
Quality, of which any Author has yet Given
us an any thing competent History. These
things I mention to You, Pyrophilus, *not*

at

at all to derogate from those Great Men; whose design seems rather to have been to deliver Principles and Summaries of Philosophy, then to insist upon Particulars; but for this purpose, that since the Nature of *Qualities* is so beneficiall a *speculation,* my labours may not be look'd upon as wholly Uselesse, though I can contribute but a little to the clearing of it: and that since 'tis so abstruse a subject, I may be pardon'd, if I sometimes misse the marke, and leave diverse things uncompleated; That being but what such great Philosophers have done before mee.

But, Pyrophylus, before I proceed to give You my Notes upon this part of our Author's Essay, that you may rightly understand my Intention in them, it will be requisite to give you three or foure Advertisements.

And first, when ever I shall speake indefinitely of *Substantiall forms,* I would alwayes be understood to except the Reasonable Soule, that is said to inform the humane Body; which Declaration I here desire may be

:   taken

The Preface.

taken notice of, once for all.

Secondly, Nor am I willing to treat of the Origine of Qualities in beasts; partly because I would not be engaged to examine, of what Nature their Soules are, and partly because it is difficult in most cases, (at least for one, that is compassionate enough,) either to make experiments upon Living animals, or to judg what influence their Life may have, upon the change of Qualities, produc'd by such Experiments.

Thirdly, The occasion of the following Reflections, being onely this, that our Author in that part of his Essay concerning Salt-peter, whereto these Notes referre, does briefly Intimate some Notions about the Natur: and Origine of Qualities; You must not exspect, that I, whose Method leads me but to Write some Notes upon this, and some other parts of this Essay, should make Solemne or Elaborate discourses concerning the Nature of particular Qualities, and that I should fully deliver my own apprehensions concerning those

B 4   Subjects

The Præface.

Subjects. For as I elsewhere sufficiently Intimate, that in these first Notes I Write as a Corpuscularian, *& set down those Things onely, that seem to have a tendency to Illustrate or Countenance the Notions or Fancies imply'd in our Author's* Essay: *So I must here Tell you, that I neither have now the Leasure, nor Pretend to the Skill, to deliver Fully the History, or to Explicate Particularly the Nature of Each several Quality.*

Fourthly, *But I consider, that the* Schools *have of late much Amus'd the World, with a way they have got, of Referring all Naturall Effects to certain Entities, that they call* Reall Qualities, *and accordingly Attribute to them a Nature distinct from the Modification of the Matter they belong to, & in some cases Separable from all Matter whatsoever, by which Meanes they have, as farre forth as their Doctrine is Acquiec'd in, made it thought Needlesse or Hopeless for men to Employ their Industry, in searching into the Nature of Particular Qualities, & their Effects. As if, (for Instance) it be Demanded,*

manded, how Snow comes to dazle the Eyes,
they will anſwer, that 'tis by a Quality of
Whiteneß that is in It; which makes all very
white Bodies produce the ſame Effeĉt; And
if You, ask what this Whiteneſs is, They will
tell you no more in ſubſtance, then that tis a
re*all* Entity, *which denominates the Parcel*
of Matter, to which it is Ioyn'd, White; & if
You further Enquire, what this real Entity,
which They call a Quality, *is, You will find,*
as Wee ſhall ſee anon, that They either Speak
of it much after the ſame rate, that They do
of their Subſtantiall Forms; (as indeed ſome
of the Modern'ſt teach, That a Quality af-
feĉts the Matter it belongs to, per modum
formæ ſecundariæ, *as they ſpeak) or at leaſt*
they will not Explicate it more Intelligibly.

And accordingly if you further Ask
them, how white Bodies in Generall do rather
Produce this effeĉt of dazling the Eyes, then
Green or Blew ones, inſtead of being told, that
the former ſort of Bodies refleĉt Outwards,
and ſo to the Eye farre more of the Incident
Light, then the Latter; You ſhall perchance
be

be told, that 'tis their respective Natures
so to act, by which way of dispatching
difficulties, they make it very easy *to solve*
All the Phænomena of Nature in Generall,
but make men think it impossible *to ex-*
plicate almost Any of them in Particular.

 And though the Unsatisfactorisneß and
Barrenneße of the School-Philosophy have
perswaded a great many Learned Men,
especially Physicians, to substitute the Chy-
mists Three principles, instead of those of
the Schools; and though I have a very
good opinion of Chymistry it self, as 'tis a
Practicall Art; yet as 'tis by Chymists
pretended to containe a Systeme of Theori-
call Principles of Philosophy, I fear it will
afford but very little satisfaction to a se-
vere enquirer, into the Nature of Quali-
ties. For besides that, as we shall more par-
ticularly see anon, there are Many Quali-
ties, which cannot with any probability
be deduc'd from Any of the three Princi-
ples; those that are ascrib'd to One, or other
of them, cannot Intelligibly be explica-
ted,

ted, without recourse to the more Compre-
hensive Principles of the Corpuscularian
Philosophy. To tell us, for instance, that
all Solidity proceeds from Salt, onely in-
forming us, (where it can plausibly be pre-
tended) in what *materiall principle or*
ingredient *that Quality* resides, *not* how
it is produced; *for this doth not teach us,*
(for example) how *Water even in exact-*
ly clos'd vessels comes to be frozen into Ice;
that is, turn'd from a fluid to a Solid Body,
without the accession of a saline ingredi-
ent (which I have not yet found pretended,
especially Glasse being held Impervious to
Salts.) Wherefore, Pyrophilus, *I thought*
it might much conduce to the understan-
ding the Nature of Qualities , To shew how
they are Generated; and by the same way, I
hop'd it might remove in some measure the
obstacle,that these Dark and Narrow Theo-
ries of the Peripateticks and Chymists may
prove to the Advancement of solid and use-
full Philosophy. That then, which I chie-
fly aime at, is to make it Probable to you by
                              Experiments,

Experiments, *(which I Think hath not yet beene done:)* *That allmost all sorts of Qualities , most of which have been by the Schooles either left Unexplicated, or Generally referr'd, to I know not what Incomprehensible Substantiall Formes*; may be produced *Mechanically ,* *'I mean by such Corporeall Agents, as do not appear, either to Work otherwise, then by vertue of the Motion, Size, Figure, and Contrivance of their own Parts, (which Attributes I call the Mechanicall Affections of Matter, because to Them men willingly Referre the various Operations of Mechanical Engines:) or to Produce the new Qualities exhibited by those Bodies their Action changes, by any other way, then by chan-*ging the Texture, or Motion, or *some other* Mechanical Affection *of the Body wrought upon. And this if I can in any Passable measure do, though but in a generall way, in some or other of each of these Three Sorts, into which the Peripateticks are wont to Divide the Qualities of Bodies, I hope I shall have*
done

done no uſeleſſe Piece of Service to Natural Philoſophy, Partly *by exciting You, and Your Learned Friends, to Enquire after more Intelligible and Satisfactory wayes of explicating Qualities*, and Partly *by* Beginning *ſuch a Collection of Materials towards the* Hiſtory *of thoſe Qualities, that I ſhall the moſt largely Inſiſt on, as Heat, Colours, Fluidity and Firmneſſe, as may invite You, and other Ingenious Men, to contribute alſo their Experiments, and Obſervations to ſo Uſefull a VVork, and thereby lay a foundation, whereon You, and perhaps I, may ſuperſtruct a more Diſtinct and Explicite Theory of Qualities, then I ſhall at preſent adventure at. And though I Know, that ſome of the things my Experiments tend to Manifeſt, may likewiſe be Confirm'd by the more obvious Phænomena of Nature, yet I Praſume You will not diſlike my Choſing to entertaine You with the Former, (though without at all Deſpiſing, or ſo much as ſtrictly forbearing to Employ the Latter,) becauſe the Changes of Qualities made by*

our

The Preface.

Our Experiments will for the most part be more Quick & Conspicuous, and the agents made use of to produce them, being of our own Applying, and oftentimes of our own Preparation, we may be thereby assisted the better to judge of what they Are, and to make an estimate of what 'tis they Do.

CONSIDE-

CONSIDERATIONS,

AND

EXPERIMENTS

touching

the Origine of Qualities, and Forms.

The Theoricall Part.

Hat before I defcend to Particulars, I may (*Pyro-philus*) furnifh you with fome General Apprehen-fion of the Doctrine (or rather the *Hypothefis*,) which is to be Collated *with*, and to be either Con-firmed, or Difproved by, the Hiftori-call Truths, that will be deliver'd con-cerning Particular Qualities, (& Forms;) I will affume the perfon of a Corpufcula-

rian,

rian, and here, at the Entrance, give you (in a general way) a brief Account of the *Hypothefis* it felfe, as it concernes the Origine of Qualities (and Forms:) and for Diftinctions fake, I fhall comprize it in the Eight following Particulars, which, that the whole Scheme may be the better Comprehended, and as it were Survey'd under one Profpect, I fhall do little more then Barely propofe *Them*, that either feem evident enough by their owne Light, or may without Præjudice have diverfe of their Proofes referv'd for proper places in the following part of this Treatife: and though there be fome *Other* Particulars, to which the Importance of the Subjects, and the Greatneffe of the (almoft Univerfall) Prejudices, that lye againft them, vvill oblige mee Immediately to annexe (for the feafonable Clearing, and Juftifying of them) fome Annotations: yet that they may, as Little as I can, Obfcure the Cohærence of the vvhole Difcourfe, as much

much of them as conveniently may be, fhall be included in [ ] Parathefes.

I. I agree with the generality of Philofophers fo far, as to allow, that there is one Catholick or Univerfal Matter common to all Bodies, by which I mean a Subftance extended, divifible and impenetrable.

II. But becaufe this Matter being in its own Nature but one, the diverfity we fee in Bodies muft neceffarily arife from fomewhat elfe, then the Matter they confift of. And fince we fee not, how there could be any change in Matter, if all its (actual or defignable) parts were perpetually at reft among themfelves, it will follow, that to difcriminate the Catholick Matter into variety of Natural Bodies, it muft have Motion in fome or all its defignable Parts: and that Motion muft have various tendencies, that which is in this part of the Matter tending one way, and that which is in that part tending another;

C   as

as we plainly fee in the Univerfe or general Mafs of Matter there is really a
great quantity of Motion, and that varioufly determin'd, and that yet diverfe
portions of Matter are at reft.

That there is Local Motion in many
parts of Matter is manifeft to fenfe, but
how Matter came by this Motion was of
Old, and is ftill hotly difputed of: for the
antient Corpufcularian Philofophers,
(whofe doctrine in moft other points,
though not in all, we are the moft incli-
nable to,)not acknowledging an Author
of the Univerfe, were thereby reduc'd
to make Motion congenite to Matter,
and confequently coëval with it; but
fince Local Motion, or an Endeavour at
it, is not included in the nature of Matter, which is as much Matter, when it
refts, as when it moves ; and fince we
fee, that the fame portion of Matter
may from Motion be reduc'd to Reft,
and after it hath continu'd at Reft, as
long as other Bodies doe not put it out

of

of that state, may by external Agents
be set a moving again; I, who am not
wont to think a man the worse Naturalist for not being an Atheist, shall not
scruple to say with an Eminent Philoso-
pher of Old, whom I find to have propos'd among the Greeks that Opinion
(for the main) that the Excellent *Des
Cartes* hath revived amongst Us, That
the Origine of Motion in Matter is from
God; and not onely so, but that thinking it very unfit to be believ'd, that
Matter barely put into Motion, and
then left to it self, should Casually constitute this beautiful and orderly
World: I think also further, that the
wise Author of Things did by establish-
ing the laws of Motion among Bodies,
and by guiding the first Motions of the
small parts of Matter, bring them to
convene after the manner requisite to
compose the World, and especially did
contrive those curious and elaborate
Engines, the bodies of living Crea-

C 2   tures,

tûres, endowing moft of them with a power of propagating their Species: But though thefe things are my Perfwa- fions, yet becaufe they are not neceffa- ry to be fuppof'd here, where I doe not pretend to deliver any compleat Dif- courfe of the Principles of Natural Phi- lophy, but onely to touch upon fuch Notions, as are requifite to explicate the Origine of Qualities and Forms, I fhall pafs on to what remains, as foon as I have taken notice, that *Local Mo- tion feems to be indeed the Principl a- mongft Second Caufes, and the Grand A- gent of all that happens in Nature:* For though Bulk, Figure, Reft, Situation, and Texture do concurre to the *Phæno- mena* of Nature, yet in comparifon of Motion they feem to be in many Cafes, Effects, and in many others, little bet- ter then *Conditions,* or *Requifites* , or Caufes *fine quibus non* , which modifie the operation, that one part of Matter by vertue of its Motion hath upon a-
nother,

nother: as in a Watch, the number, the
figure, and coaptation of the Wheels
and other parts is requifite to the fhewing the hour, and doing the other things
that may be perform'd by the Watch;
but till thefe parts be actually put into
Motion, all their other affections remaine inefficacious: and so in a Key,
though if it were too big, or too little, or
if its Shape were incongruous to that of
the cavity of the Lock, it would be unfit to be uf'd as a Key, though it were
put into Motion; yet let its bignefs and
figure be never so fit, unlefs actual Motion intervene, it will never lock or unlock any thing, as without the like a-
ctual Motion, neither a Knife nor Rafor
will actually cut, how much foever their
fhape & other Qualities may fit them to
do so. And so Brimftone, what difpofition of Parts foever it have to be turn'd
into Flame, would never be kindled,
unlefs fome actual fire, or other parcel
of vehemently and varioufly agitated

Matter should put the Sulphureous Corpuscles into a very brisk motion.

III. These two grand and most Catholick Principles of Bodies, Matter, and Motion, being thus establish'd, it will follow both, that Matter must be actually divided into Parts, that being the genuine Effect of variously determin'd Motion, and that each of the primitive Fragments, or other distinct and entire Masses of Matter must have two Attributes, its own Magnitude, or rather *Size*, and its own *Figure* or *Shape*. And since Experience shews us (especially that which is afforded us by Chymical Operations, in many of which Matter is divided into Parts, too small to be singly sensible,) that this division of Matter is frequently made into insensible Corpuscles or Particles, we may conclude, that the minutest fragments, as well as the biggest Masses of the Universal Matter are likewise endowed each with its peculiar Bulk and Shape.

Shape. For being a finite Body, its
Dimenfions muft be terminated and
meafurable: and though it may change
its Figure, yet for the fame reafon it
muft neceffarily have *fome Figure* or
other. So that now we have found out,
and muft admit three Effential Properties of each entire or undivided, though
infenfible part of Matter, namely, *Magnitude*, (by which I mean not quantity
in general, but a determin'd quantity,
which we in Englifh oftentimes call the
Size of a bodie,) *Shape*, and either *Motion* or *Reft*, (for betwixt them two
there is no mean:) the two firft of
which may be called *infeparable Accidents* of each diftinct part of Matter: *Infeparable*, becaufe being extended, and
yet finite , it is Phyfically impoffible,
that it fhould be devoid of fome Bulk
or other, and fom determinate Shape or
other; and yet *Accidents*, becaufe that
whether or no the Shape can by Phyfical Agents be alter'd or the Body fubdi-

vided,

vided,yet mentally both the one and the
other may be done, the whole essence
of Matter remaining undestroy'd.

Whether these Accidents may not
conveniently enough be call'd the
Moods or primary affections of Bodies,
to distinguish them from those lesse
simple Qualities, (as Colours, Tastes,
and Odours,) that belong to Bodies u-
pon their account, or whether with the
Epicureans they may not be called the
Conjuncts of the smallest parts of Matter, I shall not now stay to consider,
but one thing the Modern Schools are
wont to teach concerning Accidents,
which is too repugnant to our present
Doctrine, to be in this place quite omitted, namely that there are in Natural
Bodies store of *real Qualities*, and o-
ther *real Accidents*, which not onely
are no Moods of Matter, but are real
Entities distinct from it, and according
to the doctrine of many modern
Schoolmen may *exist separate* from all
Matter

Matter whatſoever. To clear this
point a little, we muſt take notice, that
Accident is among Logicians and Philoſophers uſ'd in two ſeveral ſenſes, for
ſometimes it is oppoſ'd to the 4th Præ-
dicable, (*Property*,) and is then defin'd,
" that which may be preſent or abſent,
without the deſtruction of the ſubject;
as a Man may be ſick or well , and a
Wall white or not white, and yet the
one be ſtill a Man, the other a Wall;
and this is call'd in the Schools *Accidens
prædicabile*, to diſtinguiſh it from what
they call *Accidens prædicamentale* ,
which is oppoſ'd to Subſtance: for when
things are divided by Logicians into 10
Prædicaments , or higheſt genus'es of
things, Subſtance making one of them,
all the nine other are of Accidents. And
as Subſtance is commonly defin'd to be
a thing that ſubſiſts of it ſelf, and is the
ſubject of Accidents, (or more plainly,
a real Entity or thing, that needs not
any (*created*) Being, that it may exiſt:)
ſo

ſo an Accident is ſaid commonly to be *id cujus eſſe eſt ineſſe*, and therefore *A-*
riſtotle, who uſually calls Subſtances
ſimply ὄντα, Entities, moſt commonly
calls Accidents' ὄντ⊙ ὄντα·, Entities of
Entities. Theſe needing the exiſtence
of ſome ſubſtance or other, in which
they may be, as in their ſubjeꝗ of Inhæſion. And becauſe Logicians make it
the diſcriminating note of Subſtance,
and Accident, that the former is a thing
that cannot be in another, as in its ſubjeꝗ of Inhæſion, tis requiſite to know,
that according to them, That is ſaid to
Be in a Subjeꝗ, which hath theſe three
conditions; That however it (1) *be in*
another thing, (2) *is not in it as a part*,
and (3) *cannot exiſt ſeparately* from the
thing or ſubjeꝗ, wherein it is: as a
white Wall is the ſubjeꝗ of Inhæſion
of the whiteneſs we ſee in it, which ſelf-
ſame whiteneſs, though it be not in the
wall as a part of it, yet cannot the ſelf-
ſame whiteneſs according to our Logicians

cians exift any where out of the wall,
though many other Bodies may have
the like degree of whitenefs. This premif'd, twill not be hard to difcover the
falfity of the lately mentioned Scholaftick opinion touching real Qualities
and Accidents , their doctrine about
which does, I confefs, appear to me to
be either unintelligible , or manifeftly
contradictious: for fpeaking in a Phyfical fenfe, if they will not allow thefe
Accidents to be Modes of Matter, but
Entities really diftinct from it , and in
fome cafes feparable from all Matter,
they make them indeed Accidents in
name, but reprefent them under fuch a
notion as belongs onely to Subftances;
the nature of a Subftance confifting in
this, That it can fubfift of it felfe, without being in any thing elfe, as in a fubject of Inhæfion: fo that to tell us, that
a Quality, or other Accident may fubfift without a fubject, is indeed, whatever they pleafe to call it, to allow it the
true

true Nature of Subſtance, nor will their Groundleſſe Diſtinctions do any more then keep them from ſeeming to contradict themſelves in words, whilſt Unprepoſſeſs'd perſons ſee that they do it in effect. Nor could I ever find it intelligibly made out, what theſe real Qualities may be, that they deny to be either Matter or modes of Matter, or immaterial Subſtances. When a Bowl runs along or lies ſtill, that *Motion* or *Reſt,* or *Globous figure* of the Bowl, is not *Nothing,* and yet it is not any *part* of the Bowl; whoſe whole Subſtance would remain, though it wanted which you pleaſe of theſe Accidents: and to make them *real* and *phyſical* Entities, (for we have not here to do either with *Logical* or *Metaphyſical* ones) is, as if, becauſe we may conſider the ſame Man ſitting, ſtanding, running, thirſty, hungrie, wearie, &c. we ſhould make each of theſe a diſtinct Entitie, as we do give ſome of them (as hunger, wearineſs, &c.)

diſtinct

diſtinct names. Whereas the ſubject
of all theſe Qualities is but the ſame
Man as he is conſidered with Circum-
ſtances, that make him appear different
in one caſe from what he appears in a-
nother: And it may be very uſeful to
our preſent Scope to obſerve, that not
onely diverſity of *Names*, but even diverſity of *Definitions*, doth not alwaies
infer a diverſity of *Phyſical Entities* in
the Subject, whereunto they are attributed. For it happens in many of the
Phyſical Attributes of a Body, as in
thoſe Other caſes, wherein a Man that
is a Father, a Husband, a Maſter, a
Prince, &c. may have a Peculiar Definition (ſuch as the Nature of the thing
will bear) belong unto him in each of
theſe Capacities, and yet the Man in
himſelf conſidered is but the ſame Man,
who in reſpect of differing Capacities
or Relations to other things is call'd by
differing Names, and deſcrib'd by various Definitions, which yet (as I was
saying)

faying) conclude not fo many real and diftinct Entities in the perfon fo varioufly denominated.

An
EXCURSION

about the Relative Nature *of* Phyfical Qualities.

BUt becaufe I take this Notion to be of no Small Importance towards the Avoiding of the Grand Miftake, that hath hitherto obtain'd about the Nature of Qualities, it will be worth while to · Illuftrate it a little farther. We may confider then , that when *Tubal-Cain*, or whoever elfe were the Smith, that· Invented *Locks* and *Keyes*, had made his firft Lock, (for we may Reafonably fuppofe him to have made that before the *Key*, though the Comparifon

parifon may be made ufe of withoutˊ
that Suppofition,) That was onely a
Piece of Iron , contriv'd into fuch a
Shape; and when afterwards he made a
Key to that Lock, That alfo in it felf
Confider'd, was nothing but a Piece of
Iron of fuch a Determinate Figure: but
in Regard that thefe two Pieces of Iron
might now be Applied to one another
after a Certain manner, and that there
was a Congruitie betwixt the Wards
of the Lock and thofe of the Key, the
Lock and the Key did each of them
now Obtain a new Capacity, and it became a Main part of the Notion and
Defcription of a Lock, that it was capable of being made to Lock or Unlock by that other Piece of Iron we call
a Key, and it was Lookd upon as a Peculiar Faculty and Power in the Key,
that it was Fitted to Open and Shut the
Lock, and yet by thefe new Attributes
there was not added any Real or Phyfical Entity, either to the Lock, or to the
Key,

Key, each of them remaining indeed nothing, but the same Piece of Iron, juft so Shap'd as it was before. And when our Smith made other Keyes of differing Bigneffes, or with Differing Wards, though the firft Lock was not to be open'd by any of thofe Keyes, yet that Indifpofition, however it might be Confider'd as a peculiar Power of Refi-fifting this or that Key, and might ferve to Difcriminate it fufficiently from the Locks thofe Keyes belong'd to, was nothing new in the Lock, or diftinct from the Figure it had before thofe Keyes were made. To carrie this Com-parifon a little Further, let me adde, that though one that would have De-fin'd the Firft Lock, and the Firft Key, would have Given them diftinct Defi-nitions with Reference to each other; and yet (as I was faying) thefe Defini-tions being given but upon the Score of Certain Refpects, which the Defin'd Bodies had One to Another, would not

infer,

infer, that thefe two Iron Inftruments
did Phyfically differ otherwife then in
the Figure, Size, or Contrivement of
the Iron, whereof each of them confifted. And proportionably hereunto I
do not fee, why we may not conceive,
That as to thofe Qualities (for Inftance) which we call Senfible, though by
virtue of a certain Congruity or Incon-
gruity in point of Figure or Texture,
(or other Mechanical Attributes,) to
our Senfories, the Portions of Matter
they Modifie are enabled to produce
various Effects, upon whofe account we
make Bodies to be Endow'd with Qualities; yet They are not in the Bodies
that are Endow'd with them any Real
or Diftinct Entities, or differing from
the Matter its felf, furnifh'd with fuch
a Determinate Bignefs, Shape, or o-
ther Mechanical Modifications. Thus
though the modern Gold-Smiths and
Refiners reckon amongft the moft di-
ftinguifhing Qualities of Gold, by which

D   men

men may be certain of its being True,
and not Sophifticated, that is eafily diffoluble in *Aqua Regis*, and that *Aqua
Fortis* will not work upon it ; yet thefe
Attributes are not in the Gold any thing
diftinct from its peculiar Texture, nor
is the Gold we have now of any other
Nature , then it was in *Pliny's* time ,
when *Aqua Fortis* and *Aqua Regis* had
not been Found out, (at leaft in thefe
parts of the World,) and were utterly
unknown to the Roman Gold-Smiths.
And this Example I have the rather
pitch'd upon, becaufe it affords me an
Opportunity to reprefent, that, unlefs
we admit the Doctrine I have been Propofing, we muft Admit, that a Body
may have an almoft Infinite Number of
New Real Entities accruing to it, without the Intervention of any Phyfical
Change in the Body its felf. As for
Example, Gold was the fame Natural
Body immediately before *Aqua Regis*
and *Aqua Fortis* were firft made, as it
was

(21)

was immediately after, and yet now 'tis reckon'd amongſt its Principal Properties, that it is diſſoluble by the Former of thoſe two Menſtruums, and that it is not like other Mettals Diſſoluble or Corrodible by the Latter. And if one ſhould Invent another Menſtruum, (as poſſibly I may Think my ſelf Maſter of ſuch a one) that will but in part diſſolve pure Gold, and change ſome part of it into another Metalline Body, there will then ariſe another new Property, whereby to diſtinguiſh That from other Mettals; and yet the Nature of Gold is not a whit other now, then it was before this laſt Menſtruum was firſt made. There are ſome Bodies not Cathartick, nor Sudorifick, with ſome of which Gold being joyn'd acquires a Purgative Vertue, and with others a power to procure Sweat; and in a word, Nature her ſelf doth, ſometimes otherwiſe, and ſometimes by Chance, produce ſo many things, that have new Relations unto o

D 2   thers:

thers: And Art, especially assisted by
Chymistry, may, by variously dissipa-
ting Natural Bodies, or Compounding either them, or their Constituent
Parts with one another, make such an
Innumerable Company of new Productions, that will each of Them have new
operations, either immediately upon our
Sensories, or upon other Bodies, whose
Changes we are able to perceive, that
no man can know, but that the most
Familiar Bodies may have Multitudes
of Qualities, that he dreams not of, and
a Considering man will hardly imagine,
that so numerous a Croud of real Physical Entities can accrue to a Body,
whilst in the Judgment of all our Senses it remains Unchang'd, and the Same
that 'twas before.

 To clear this a little farther, we may
adde, that beaten Glass is commonly
reckon'd among Poisons; and (to skip
what is mention'd out of *Sanctorius*, of
the Dysentery procur'd by the Fragments

ments of it) I remember *Cardan* hath
a ſtory, That in a Cloiſter, where he
had a Patient then like to die of tor-
ments in the Stomach, two other Nuns
had been already kill'd by a diſtracted
Woman, that having Caſually got Free,
had mixt beaten Glaſs with Peaſe, that
were eaten by theſe three, and diverſe
others of the Siſters (who yet eſcap'd
unharm'd.) Now though the powers of
Poiſons be not onely look'd upon as
real Qualities, but are reckoned among
the *Abſtruſeſt* ones: yet this Deleterious
Faculty, which is ſuppoſ'd to be a Peculiar and Superadded Entitie in the
beaten Glaſſe, is really nothing diſtinct
from the Glaſs its ſelf, (which though
a Concrete made up of thoſe Innocent
Ingredients, Salt and Aſhes, is yet a
hard and ſtiffe Body,) as it is furniſh'd
with that determinate Bigneſs, and Figure of Parts, which have been acquir'd

* *Cardan: Contradict.9.lib.2.Tract.5.a pud Schenckium.*

by

by Comminution. For thefe Glaffy
Fragments being many, and Rigid, and
fomewhat Small, (without yet being
fo fmall as Duft,) and endow'd with
fharp Points and cutting Edges, are enabled by thefe Mechanical Affections to
Pierce or Wound the tender Membranes of the Stomach and Guts, and
cut the flender Veffels that they meet
with there, whereby naturally enfue
great Gripings and Contorfions of the
injur'd Parts, and oftentimes Bloudy
Fluxes occafion'd by the perforation of
the Capillary Arteries, and the great
irritation of the Expulfive Faculty, and
fometimes alfo not onely horrid Con-
vulfions by Confent of the Brain and
Cerebellum, with fome of the Nervous
or Membranous parts that happen to be
hurt, but alfo Dropfies occafioned by
the great lofs of Bloud we were juft
now fpeaking of. And it agrees very
well with this Conjecture, that beaten
Glafs hath diverfe times been obferv'd
to

to have done no Mifchief to Animals
that have fwallowed it: For there is no
Reafon it fhould, in cafe the Corpufcles
of the Powder either chance to be fo
fmall, as not to be fit to wound the
Guts, which are ufually lin'd with a flimy fubftance, wherein very minute
Powders may be as it vvere fheath'd,
and by that means hinder'd from hurt-
ing the Guts, (infomuch that a fragment of Glafs vvith three very fharp
corners, hath been obferv'd to have for
above eighteen Months lain * inoffenfive even in a nervous and very fenfible
part of the body,) out of vvhich they
may with the groffer Excrements of the
Lower Belly be harmelefly Excluded,

---

* This memorable Accident happen'd to a Senator
of Berne, who was cur'd by the Experienc'd Fabricius
Hildanus, that gives a long Account of it to the Lear-
ned Horftius, among whofe Obfervations tis extant:
(Lib. 2. obferv. 3 5.) who afcribes the Indolence of the
Part,whilft uncomprefs'd, to fome flimy Juice, (fami-
liar enough to thofe Tendinous parts,) wherein the
Glaffy fragment was as it were Bedded.

D 4   efpe-

efpecially in fome Individuals, whofe
Guts and Stomach too may be of a
much ftronger Texture, and better
Lin'd or Stuff'd with Grofs and Slimy
Matter, then thofe of others. And accordingly we fee, that the Fragments
of Saphires, Chriftals, and ev'n Rubies,
which are much harder then Glafs; are
innocently, though perhaps not very
effectually us'd by Phyficians, (and I
have feveral times taken That without
Inconvenience) in Cordial Compofiti-
ons, becaufe of their being by Grinding
reduc'd to a Powder too Subtle to Ex-
coriate, or Grate upon the Stomach, or
Guts; and probably 'twas upon fome
fuch Account, that That happen'd
which is related by *Cardan* in the fame
place, namely, That though the three
Nuns we have been fpeaking of were
Poifon'd by the Glafs, yet many others
who eat of the other Portions of the
fame mingled Peafe, receiv'd no mifchief thereby. (But of this fubject
more

more † elſewhere.)

And this puts me in mind to adde, That the Multiplicity of Qualities, that are ſometimes to be met with in the ſame Natural Bodies, needs not make men rejeɕt the Opinion we have been propoſing, by perſwading them, that ſo many Differing Attributes, as may be ſometimes found in one and the ſame Natural Body, cannot proceed from the bare Texture, and other Mechanical Affeɕtions of its Matter. For we muſt conſider each Body, not barely as it is in it ſelf an entire and diſtinɕt portion of Matter, but as it is a Part of the Univerſe, and conſequently plac'd among a great Number and Variety of other Bodies, upon which it may Aɕt, and by which it may be aɕted on, in many waies, (or upon many Accounts,) each of which Men are wont to Fancy,

---

† In thoſe Notes about Occult Qualities, where the Deleterious Faculty attributed to Diamonds is conſidered.

as a diftinct Power or Quality in the Body, by which thofe Actions, or in which thofe Paffions are produc'd. For if we thus confider Things, we fhall not much wonder, that a Portion of Matter, that is indeed endow'd but with a very few Mechanical Affections, as fuch a determinate Texture and Motion, but is plac'd among a multitude of other Bodies, that differ in thofe Attributes from it, and one another, fhould be capable of having a great Number and Variety of Relations to thofe other Bodies, and confequently fhould be thought to have many Diftinct Inhærent Qualities, by fuch as look upon thofe feveral Relations or Refpects it may have to Bodies without it, as Real and Diftinct Entities implanted in the Body it felf. When a Curious Watch is going, though the Spring be that which puts all the Parts into Motion, yet we do not Fancie (as an *Indian* or *Chinois* would perchance do) in this

Spring

Spring one Faculty to move the Index uniformely round the Dial-plate, another to ſtrike the Hour, and perhaps a Third to give an Alarme, or ſhew the Age of the Moon, or the Tides; all the action of the Spring, (which is but a flexible piece of Steel, forcibly coil'd together,) being but an Endeavour to dilate or unbind its ſelf, and the reſt being perform'd by the various Reſpects it hath to the ſeveral Bodies (that compoſe the Watch) among which it is plac'd, and which they have One to another. We all know, that the Sun hath a power to Harden Clay, and Soften Wax, and Melt Butter, and Thaw Ice, and turn Water into Vapours, and make Air expand it ſelf in Weather-Glaſſes, and contribute to Blanch Linnen, and make the White skin of the Face Swarthy, and Mowed Graſs Yellow, and ripen Fruit, hatch the Eggs of Silk-worms, Caterpillars, and the like Inſects, and perform I know not how

many

many other things, divers of which feem
contrary Effects, and yet thefe are not
diftinct Powers or Faculties in the Sun,
but onely the Productions of its Heat,
(which it felf is but the brisk, and con-
fuf'd Local Motion of the Minute parts
of a Body,) diverfify'd by the differing
Textures of the Body that it chances to
work upon, and the Condition of the o-
ther Bodies that are concern'd in the
Operation. And therefore whether
the Sun in fome cafes have any Influence at all diftinct from its Light and
Heat, we fee, that all thofe *Phænomena*
we have thought fit to name are producible by the heat of the common Culi-
nary Fire duly apply'd and regulated.
And fo, to give an *Inftance* of another
Kind, when fome years fince, to Try
fome Experiments about the Propaga-
tion of Motion, with Bodies lefs capable of being batter'd by one another,
then thofe that have been formerly imploy'd; I caus'd fome folid Bals of Iron
skil-

skilfully harden'd , and exquifitely
fhap'd and glaz'd , to be purpofely
made; each of thefe polifhed Balls was a
Sphærical Looking-Glafs, which plac'd
in the mid'ft of a Room, would exhibit
the Images of the Objects round about
it, in a very regular and pleafing Perfpective. It would Contract the Image,
and Reflect the Beams of the Sun, after
a manner differing from Flat and from
Convex Looking Glaffes. It would in
a neat Perfpective leffen the Image of
him that look'd upon it; and bend it,
and it would fhew that Image, as if it
were behind the Surface, and within the
folid fubftance of the Sphære, and in
fome it had all thofe Diftinct, and fome
of them wonderful Properties , which
either Antient or Modern Writers of
Catoptricks have demonftrated to belong to Sphærical *Specula*, as fuch: and
yet the Globe furnifh'd with all thefe
Properties and Affections, was but the
Iron it felf reduc'd by the Artificer to
a

a Sphærical Figure, (for the Glafs, that made it Specular, was not diftinct from the Superficial parts of the Iron, re- duc'd all of them to a Phyfically equal diftance from the Center.) And of *specula*, Sphærical enough as to fenfe, you may make ftore in a trice, by break- ing a large Drop of Quick-filver into feveral little ones, each of which will ferve for Objects plac'd pretty near it, and the fmaller of which (being the leaft deprefs'd in the middle by their own weight, and confequently more perfectly Globous,) may with a good Microfcope plac'd in a Window afford you no unpleafant profpect of the neigh- bouring Objects, and yet to reduce a parcel of Stagnant Quickfilver, which will much æmulate a Flat Looking- Glafs, into many of thefe little Sphæri- cal *Specula*, whofe Properties are fo dif- fering from thofe of Plain ones, there intervenes nothing but a fleight Local Motion, which in the twinckling of an

Eye

Eye changeth the Figure of the felf fame Matter.

I have faid thus much (*Pyrophilus*) to remove the Miftake, That *every thing men are wont to call a Quality*, muft needs be a Real and Phyfical Entity, becaufe of the Importance of the Subjeɑ; and yet I have omitted fome things that might have been pertinently added, partly becaufe I may hereafter have Opportunity to take them in, and partly becaufe I would not any farther lengthen this Excurfion, which yet I muft not Conclude, till I have added this fhort Advertifement.

That I have chofen to Declare what I mean by Qualities, rather by Examples, then Definitions, *partly* becaufe being immediately or reduɑively the Objeɑs of fenfe, Men generally underftand pretty well what one another mean, when they are fpoken of: (As to fay, that the Taft of fuch a thing is Saline or Sowr, or that fuch a Sound is

Melo-

Melodious, Shrill, or Jarring, (especially
if when we speak of Sensible Qualities,
we adde some Enumeration of particular Subjects, wherein they do the most
Eminently reside,) will make a Man as
soon understood, as if he should go a-
bout to give Logical Definitions of
those Qualities:) and *partly* because the
Notions of things are not yet so well
stated, and agreed on, but that it is many times difficult to Assign their true
Genus's: and *Aristotle* himself doth not
onely define *Accidents* without setting
down their *Genus*, but when he comes
to define *Qualities*, he tels us, that *Quality is that by which a thing is said to be
Qualis*, where I would have you take
notice both, that in his Definition he o-
mits the *Genus*, and that 'tis no such easy Thing to give a very good Definition of Qualities, since he that is reputed
the great Master of Logick, where he
pretends to give us one, doth but upon
the matter define the thing by the same
thing,

thing; for 'tis suppos'd to be as little known what *Qualis* is, as what *Qualitas* is, and me thinks he does just as if I should define Whiteness to be that, for which a thing is called White, or Vertue, that for which a Man is said to be Vertuous †. Besides that, I much

---

† Since the writing of this, the Author found, that some of the Eminentest of the modern Schoolmen themselves, have been, as well as he, unsatisfied with the Aristotelian Definition of Quality: concerning which (not to mention *Revius*, a Learned Protestant Annotator upon *Suarez*.) *Ariaga* sayes (*disp. 5. sect. 2. subs. 1.*) *Per hanc nihil explicatur; nam de hoc quærimus, quid sit esse qual., dices habere qualitatem ; bonus Circulus: qualitas est id quo quis fit qualis, & esse qualem est habere Qualitatem.* And even the famous Jesuit *Suarez*, though he endeavours to excuse it, yet confesseth, that it leaves the proper Notion of Quality as obscure to us as before: (*Quæ definitio, saith he, licèt ea ratione essentialis videatur, quod detur per habitudinem ad effectum formalem, quem omnis Forma essentialiter respicit, tamen quod ad nos spectat, aquè obscura nobis manet propria ratio Qualitatis.*) *Suarez Disputat. Metaphysic. 42.* But *Hurtadus* (in his Metaphysical Disputations) speaks more boldly, telling us roundly, that it is *Non tam Definitio, quàm inanis quædam Nugatio*, which makes me the more wonder, that a famous Cartesian (whom I forbear to name) should content himself to give us such an Insignificant, or at least Superficial Definition of Quality.

E   doubt,

doubt, whether his Definition be not Untrue as well as Obscure, for to the Question, *Qualis res est?* Answer may be return'd out of *some*, if not *all* of the other *Prædicaments of Accidents*: which some of the Modern Logicians being aware of, they have endeavoured to salve the matter with certain Cautions and Limitations, which however they may argue the Devisors to be ingenious, do, for ought I can discern, leave us still to seek for a right and intelligible Definition of Quality in general, though to give such a one be probably a much easier Task, then to define many Qualities, that may be nam'd in particular, as Saltness, Sowrness, Green, Blew, and many others, which when we hear nam'd, every man knows what is meant by them, though no man (that I know of) hath been able to give accurate Definitions of them.

IV. And if we should conceive, that
all

all the reſt of the Univerſe were annihilated, except any of theſe entire and
undivided Corpuſcles, (treated of in
the 3ᵈ Particular foregoing,) it is hard
to ſay what could be attributed to it, beſides Matter, Motion (or Reſt,) Bulk,
and Shape, (whence by the way you
may take notice, that Bulk, though u-
ſually taken in a Comparative ſenſe, is
in our ſenſe an abſolute Thing, ſince
a Body would have it, though there
were no other in the World.) But now
there being actually in the Univerſe
great Multitudes of Corpuſcles mingled among themſelves, there ariſe in
any diſtinct portion of Matter, which a
number of them make up, two new
Accidents or Events: the one doth
more relate to each particular Corpuſ-
cle in reference to the (really or ſuppo-
ſedly) ſtable Bodies about it, namely
its *Poſture*; (whether Erected, Inclin'd,
or Horizontal:) And, when two or
more of ſuch Bodies are plac'd one by
another,

anorher, the manner of their being fo
plac'd, as one befides another, or one
behind another, may be call'd their
Order; as I remember, *Ariftotle* in his
Metaphyficks, *lib. 1.cap. 4.* recites this
Example out of the antient Corpufcularians, That *A and N differ in Figure,
and A N and N A in Order, Z and N
in Scituation*: and indeed Pofture and
Order feem both of them reducible to
Scituation.   And when many Corpufcles do fo convene together as to compofe any diftinct Body, as a Stone, or a
Mettal, then from their other Accidents
(or Modes,) and from thefe two laft
mention'd, there doth emerge a certain
Difpofition or Contrivance of Parts in
the whole, which we may call the *Texture* of it.

 V. And if we fhould conceive all
the reft of the Univerfe to be annihilated, fave one fuch Body, fuppofe a Mettal or a Stone, it were hard to fhew, that
there is Phyfically any thing more in it
           then

then Matter, and the Accidents we have
already named. But now we are to
confider, that there are *de facto* in the
world certain fenfible and rational Beings, that we call Men, and the body of
Man having feveral of its external parts,
as the Eye, the Ear, &c. each of a diftinct and peculiar Texture, whereby it
is capable to receive Impreffions from
the Bodies about it, and upon that account it is call'd an Organ of Senfe, we
muft confider, I fay, that thefe Senfories may be wrought upon by the Figure, Shape, Motion, and Texture of
Bodies without them, after feveral
waies, fome of thofe External Bodies
being fitted to affect the Eye, others the
Ear, others the Noftrils, &c. And to
thefe Operations of the Objects on the
Senfories, the Mind of Man, which upon the account of its Union with the
Body perceives them, giveth diftinct
Names, calling the one Light or Colour, the other Sound, the other Odour,

&c.

&c. And becaufe alfo each Organ of
Senfe, as the Eye, or the Palat, may
be it felf differingly affected by External Objects, the Mind likewife gives
the Objects of the fame Senfe diftinct
Appellations, calling one colour Green,
the other Blew, and one taft Sweet, and
another Bitter, &c. Whence Men have
been induc'd to frame a long Catalogue
of fuch Things as, for their relating to
our Senfes, we call Senfible Qualities;
and becaufe we have been converfant
with them, before we had the ufe of
Reafon, and the Mind of Man is prone
to conceive almoft every Thing (nay
even Privations, as Blindnefs, Death,
&c.) under the notion of a true Entitie
or Subftance as it felf is; we have been
from our Infancy apt to imagine, that
thefe Senfible Qualities are Real Beings, in the Objects they denominate,
and have the faculty or power to work
fuch and fuch things; as Gravity hath a
power to ftop the motion of a Bullet
shot

ſhot upwards , and carry that ſolid
Globe of Matter toward the Center of
the Earth, whereas indeed (according
to what we have largely ſhewn above)
there is in the Body , to which theſe
Senſible Qualities are attributed, nothing of Real and Phyſical, but the Size,
Shape, and Motion, or Reſt of its component Particles, together with that
Texture of the whole, which reſults
from their being ſo contriv'd as they
are; nor is it neceſſary they ſhould have
in them any thing more, like to the
Ideas they occaſion in us, thoſe Ideas
being *either* the Effects of our Præju-
dices, or Inconſideratenefs, *or* elſe to be
fetcht from the Relation, that happens
to be betwixt thoſe Primary Accidents
of the Senſible Object, and the peculiar
Texture of the Organ it affects; as when
a Pin, being run into my Finger, cauſeth
pain, there is no diſtinct Quality in the
Pin anſwerable to what I am apt to fancie Pain to be, but the Pin in it ſelf is

E 4   one-

onely flender, ftiff, and fharp, and by
thofe qualities happens to make a Solution of Continuity in my Organ of
Touching, upon which , by reafon of
the Fabrick of the Body, and the intimate Union of the Soul with it, there a-
rifeth that troublefome kind of Perception which we call Pain, and I fhall anon
more particularly fhew, how much that
depends upon the peculiar fabrick of
the Body.

VI. But here I forefee a Difficulty,
which being perhaps the chiefeft, that
we fhall meet with againft the Corpufcular Hypothefis , it will deferve to
be, before we proceed any farther, taken
notice of. And it is this, that, whereas
we explicate Colours, Odours, and the
like fenfible Qualities by a *relation to
our Senfes*, it feems evident, that they
have an *abfolute* Being irrelative to *Us*;
for, Snow (for inftance) would be white,
and a glowing Coal would be hot,
though there were no Man or any other

Animal

Animal in the World: and 'tis plain,
that Bodies do not onely by their Qualities work upon *Our fenfes* , but upon
other, and thofe, Inanimate *Bodies*; as
the Coal will not onely heat or burn a
Man's hand if he touch it, but would
likewife heat Wax, (even fo much as to
melt it, and make it flow,) and thaw Ice
into Water, though all the Men, and
fenfitive Beings in the World were annihilated. To clear this Difficulty, I
have feveral things to reprefent, and,

1. I fay not, that there are no other
Accidents in Bodies then Colours, O-
dours, and the like; for I have already
taught, that there are fimpler and more
Primitive Affections of Matter, from
which thefe Secondary Qualities, if I
may fo call them, do depend: and that
the Operations of Bodies upon one another fpring from the fame, we fhall fee
by and by.

2. Nor do I fay, that all Qualities of
Bodies are *directly Senfible*; but I obferve,

ſerve, that when one Body works upon another, the knowledg we have of their Operation, proceeds, either from ſome ſenſible Quality, or ſome more Catholick affection of Matter, as Motion, Reſt, or Texture, generated or deſtroy'd in one of them; for elſe it is hard to conceive, how we ſhould come to diſcover what paſſes betwixt them.

3. We muſt not look upon every diſtinct Body, that works upon our Senſes, as a bare lump of Matter of that bigneſs and outward ſhape, that it appears of; many of them having their parts curiouſly contriv'd, and moſt of them perhaps in motion too. Nor muſt we look upon the Univerſe that ſurrounds us, as upon a moveleſs and undiſtinguiſh'd Heap of Matter, but as upon a great Engine, which, having either no Vacuity, or none that is conſiderable, betwixt its parts (known to us,) the actions of particular Bodies upon one another muſt not be barely æſtimated,

mated, as if two Portions of Matter of
their Bulk and Figure were plac'd in
some imaginary Space beyond the
World, but as being scituated in the
World, constituted as it now is, and
consequently as having their action upon each other liable to be promoted,
or hindred, or modify'd by the Actions of other Bodies besides them: as in
a Clock, a small force apply'd to move
the Index to the Figure of 12, will make
the Hammer strike often and forcibly a-
gainst the Bell, and will make a far greater Commotion among the Wheels
and Weights, then a far greater force
would do, if the Texture and Contrivance of the Clock did not abundantly
contribute to the Production of so great
an Effect. And in agitating Water into Froth, the Whitenefs would never
be produc'd by that Motion, were it not
that the Sun, or other Lucid Body, shining upon that Aggregate of small
Bubbles, enables them to reflect confufedly

fedly great store of little, and as it were contiguous lucid images to the Eye. And so the giving to a large Metalline Speculum a Concave figure, would never enable it to set Wood on fire, and even to melt down Mettals readily, if the Sun beams, that in Cloudless dayes do, as to sense, fill the Air, were not by the help of that Concavity, thrown together to a Point. And to shew You by an eminent Instance, how various and how differing Effects the Same action of a Natural Agent may produce, according to the several Dispositions of the Bodies it works upon, do but consider, that in two Eggs, the one Prolifick, the other Barren, the sense can perhaps distinguish before Incubation no difference at all; and yet these Bodies, outwardly so like, do so differ in the internal disposition of their parts, that if they be both exposs'd to the same degree of Heat, (whether of a Hen, or an Artificial Oven,) that Heat will change the one into a putrid

trid and ſtinking Subſtance, and the o-
ther into a Chick, furniſh'd with great
variety of Organical parts of very differing conſiſtences, and curious as well as
differing Textures.

4. I do not deny, but that Bodies
may be ſaid, in a very favourable ſenſe,
to have thoſe Qualities we call Senſible,
though there were no Animals in the
World: for a Body in that caſe may differ from thoſe Bodies, which now are
quite devoid of Quality, in its having
ſuch a diſpoſition of its Conſtituent
Corpuſcles, that in caſe it were duely
apply'd to the Senſory of an Animal,
it would produce ſuch a ſenſible Quality, which a Body of another Texture
would not; as though if there were no
Animals, there would be no ſuch thing
as Pain, yet a Pin may upon the account
of its Figure be fitted to cauſe pain, in
caſe it were mov'd againſt a Man's finger; whereas a Bullet, or other blunt
Body mov'd againſt it with no greater
force,

force, will not cause any such perception of pain. And thus Snow, though if
there were no Lucid Body nor Organ of
Sight in the World, it would exhibit
no Colour at all, (for I could not find
it had any in places exactly darkned,)
yet it hath a greater disposition then
a Coal or Soot to reflect store of Light
outwards, when the Sun shines upon
them all three. And so we say, that a
Lute is in tune, whether it be actually
plaid upon or no, if the Strings be all
so duly stretcht, as that it would appear
to be in Tune, if it were play'd upon.
But as if You should thrust a Pin into
a man's Finger, both a while before and
after his Death, though the Pin be as
sharp at one time as at another, and ma-
keth in both cases alike a Solution of
Continuity; yet in the former case, the
Action of the Pin will produce Pain, and
not in the latter, because in this the
prick'd Body wants the Soule, and consequently the Perceptive Faculty: so
if

if there were no Senfitive Beings, thofe Bodies that are now the Objects of our Senfes, would be but *difpofitively*, if I may fo fpeak, endow'd with Colours, Tafts, and the like ; and *actually* but onely with thofe more Catholick Affections of Bodies, Figure, Motion, Texture, &c.

To illuftrate this yet a little farther, fuppofe a Man fhould beat a Drum at fome diftance from the mouth of a Cave, conveniently fcituated to return the Noife he makes; although Men will prefently conclude, that That Cave hath an Echo, and will be apt to fancy upon that account fome Real Property in the place, to which the Echo is faid to belong, and although indeed the fame Noife made in many other of the neighbouring places, would not be reflected to the Eare, and confequently would manifeft thofe places to have no Echos; yet to fpeak Phyfically of things, this Peculiar Quality or Property

perty

perty we fancy in the Cave, is in It nothing elfe but the Hollownefs of its Figure, whereby 'tis fo difpof'd, as when the Air beats againft it, to reflect the Motion towards the place whence that Motion began; and that which paffeth on this occafion is indeed but this, That the Drum ftick falling upon the Drum, makes a Percuffion of the Air, and puts that Fluid Body into an Undulating Motion, and the Aery Waves thrufting on one another, 'till they arrive at the hollow Superficies of the Cave, have by reafon of its refiftance and figure, their Motion determin'd the contrary way, namely backwards towards that part where the Drum was, vvhen it vvas ftruck; fo that in That, vvhich here happens, there intervenes nothing but the Figure of one Body, and the Motion of another, though if a Man's Ear chance to be in the way of thefe Motions of the Air forwards and backvvards, it gives him a Perception of them,

them, which he calls Sound; and because these Perceptions, which are suppos'd to proceed from the same percuf-
sion of the Drum, and thereby of the
Air, are made at distinct times one after
another , That hollow Body , from
whence the Last Sound is conceiv'd to
come to the Air, is imagin'd to have a
peculiar Faculty, upon whose account
Men are wont to say, that such a place
hath an Echo.

5. And whereas one Body doth often
seem to produce in another divers such
Qualities, as we call Sensible , which
Qualities therefore seem not to need
any reference to our Senses, I consider,
that when one Inanimate Body works
upon another, there is nothing really
produc'd by the Agent in the Patient,
save some Local Motion of its Parts, or
some Change of Texture consequent
upon that Motion; and so, if the Patient come to have any sensible Quality,
that it had not before, it acquires it up-

on

on the fame account, upon which other Bodies have it, and it is but a confequent to this Mechanical Change of Texture, that by means of its Effects upon our Organs of Senfe, we are induc'd to attribute this or that fenfible Quality to it. *As* in cafe a Pin fhould chance by fome inanimate Body to be driven againft a Man's Finger, that which the Agent doth, is but to put a fharp and flender Body into fuch a kind of Motion, and that which the Pin doth, is to pierce into a Body that it meets with, not hard enough to refift its Motion, and fo that upon this there fhould enfue fuch a thing as Pain, is but a Confequent, that fuperadds nothing of Real to the Pin that occafions that Pain. *So* if a piece of Tranfparent Ice be, by the falling of fome heavy and hard Body upon it, broken into a Grofs Powder that looks Whitifh, the falling Body doth nothing to the Ice but break it into very fmall Fragments, lying confufedly upon one another,

another; though by reafon of the Fabrick of the World, and of our Eyes,
there doth in the day time upon this
Comminution , enfue fuch a kind of
copious Reflection of the incident Light
to our Eyes, as we call Whiteneffe: and
when the Sun, by thawing this broken
Ice, deftroyes the Whitenefs of that
portion of Matter, and makes it become
Diaphanous, which it was not before,
it doth no more then alter the Texture
of the Component parts, by putting
them into Motion, and thereby into a
new Order; in which, by reafon of the
difpofition of the Pores intercepted betwixt them, they reflect but few of the
incident beams of Light , and tranfmit
moft of them. Thus when with a Bur-
nifher You polifh a rough piece of Silver, that which is really done, is but the
Depreffion of the little Protuberant
parts into one Level with the reft of
the Superficies; though upon this Mechanical change of the Texture of the

Super-

Superficial parts, we Men say, that it
hath loft the Quality of Roughnefs,
and acquir'd that of Smoothnefs, becaufe that whereas before, the little Ex-
ftancies by their Figure refifted a little
the Motion of our Finger, and grated
upon them a little, our Fingers now
meet with no fuch offenfive Refiftance.
'Tis true that the Fire doth thaw Ice,
and alfo both make Wax flow, and enable it to burn a Man's hand, and yet
this doth not neceffarily argue in it any
Inhærent Quality of Heat, diftinct from
the Power it hath of putting the fmall
parts of the Wax into fuch a Motion,
as that their Agitation furmounts their
Cohæfion ; which Motion, together
with their Gravity, is enough to make
them *pro tempore* conftitute a Fluid Body: and *Aqua Fortis*, without any (fenfible) Heat, will make Camphire, caft
on it, affume the form of a Liquor diftinct from it; as I have try'd , that a
ftrong Fire will alfo make Camphire
                                      fluid:

fluid: not to adde, that I know a Liquor,
into which certain Bodies being put,
when both it Self, (as well as They,)
is *actually cold*, (and confequently when
You would not fufpect it of an Actual
Inhærent Heat) will not onely fpeedily diffipate many of their parts into
Smoak, but leave the reft Black, and
burnt almoft like a Coal. So that though
we fuppofe the Fire to do no more then
varioufly and briskly to agitate the Infenfible parts of the Wax, That may
fuffice to make us think the Wax endow'd with a Quality of Heat: becaufe
if fuch an Agitation be greater then that
of the Spirit, and other parts of our Organs of Touching, That is enough to
produce in us that Senfation we call
Heat; which is fo much a Relative to
the Senfory which apprehends it, that
vve fee, that the fame Lukevvarm Water, that is, vvhofe Corpufcles are moderately agitated by the Fire, will appear
hot to one of a Man's hands, if That be

very cold; and cold to the other, in
cafe it be very hot, though both of
them be the fame Man's hands. To be
fhort, if we fancy any two of the Bodies
about us, as a Stone, a Mettal, &c. to
have nothing at all to do with any other
Body in the Univerfe, 'tis not eafy to
conceive, either how one can act upon
the other, but by Local Motion (of the
whole Body, or its Corporeal Effluvia;)
or how by Motion it can do any more,
then put the Parts of the other Body
into Motion too, and thereby produce
in them a Change of Scituation and
Texture, or of fome other of its Mechanical Affections: though this (Paf-
five) Body being plac'd among other
Bodies in a World conftituted as ours
now is, and being brought to act upon
the moft curioufly contriv'd Senfories
of Animals, may upon both thefe accounts exhibit many differing fenfible
Phænomena; which however we look
upon them as diftinct Qualities, are consequently

fequently but the Effects of the often
mention'd Catholick affections of Matter, and deducible from the Size, Shape,
Motion (or Reft,) Pofture, Order, and
the refulting Texture of the Infenfible
parts of Bodies. And therefore though,
for fhortnefs of fpeech, I fhall not fcruple to make ufe of the word *Qualities*,
fince it is already fo generally receiv'd,
yet I would be underftood to mean
them in a fenfe fuitable to the Doctrine
above deliver'd. As if I fhould fay, that
Roughneffe is apt to grate and offend
the Skin, I fhould mean, that a File or
other Body, by having upon its Surface
a multitude of little hard and exftant
Parts, and of an Angular or fharp Figure, is qualify'd to work the mention'd
Effect: and fo if I fhould fay, that Heat
melts Mettals, I fhould mean, that this
Fufion is effected by Fire, or fome other
Body, which by the various and vehement Motion of its infenfible parts,
does to us appear Hot. And hence,

F 4   (by

(by the way,) I prefume You will eafily guefs at what I think of the Contro-
verfy fo hotly difputed of late betwixt
two parties of Learned Men, whereof
the One would have all Accidents to
worke onely in virtue of the Matter
they refide in, and the Other would
have the Matter to act onely in virtue
of its Accidents: for confidering, that
on the one fide, the Qualities, we here
fpeak of, do fo depend upon Matter,
that they cannot fo much as have a Being but in, and by it; and on the other
fide, if all Matter were but quite devoid
of Motion, (to name now no other Accidents,) I do not readily conceive, how
it could operate at all, I think it is fafeft
to conclude, That neither Matter, nor
Qualities apart, but both of them con-
jointly do perform, what we fee done
by Bodies to one another, according to
the Doctrine of Qualities juft now deliver'd.

of

(*Of the Nature of a* Forme.)

VII. VV E may now advance fomewhat farther, and confi-
der, that Men having taken notice, that
certain confpicuous Accidents were to
be found affociated in fome Bodies, and
other Conventions of Accidents in o-
ther Bodies, they did for conveniency,
and for the more expeditious Expreffion of their Conceptions agree to diftinguifh them into feveral Sorts, which
they call *Genders* or *Species*, according
as they referr'd them either upwards to
a more Comprehenfive fort of Bodies,
or downward to a narrower Species, or
to Individuals: As, obferving many Bodies to agree in being Fufible, Malleable, Heavy, and the like, they gave to
that fort of Body the name of *Mettal*,
which is a *Genus* in reference to Gold,
Silver, Lead, and but a *Species* in reference to that fort of mixt Bodies they
call

call *Foßilia.* This *superior Genus* com-
prehending both Mettals, Stones, and
diverse other Concretions, though it
self be but a *Species* in respect of Mixt
Bodies. Now when any Body is referr'd
to any particular *species,* (as of a Mettal, a Stone, or the like,) because Men
have for their Convenience agreed to
signifie all the Essentials requisite to
constitute such a Body by one Name,
most of the Writers of Physicks have
been apt to think, that besides the common Matter of all Bodies, there is but
One thing that discriminates it from o-
ther Kinds, and makes it what it is, and
this for brevities sake they call a *Forme;*
which, because all the Qualities and o-
ther Accidents of the Body must depend on it, they also imagine to be a very Substance, and indeed a kind of
Soule, which united to the gross Matter composes with it a Natural Body,
and acts in it by the several Qualities to
be found therein, which Men are wont
to

to afcribe to the Creature fo compof'd,
But as to this affair, I obferve, that if
(for Inftance) You ask a Man, what
Gold is, if he cannot fhew you a piece
of Gold, and tell You, This is Gold,
he will defcribe it to You as a Body,
that is extremely Ponderous, very Malleable and Ductile, Fufible and yet Fixt
in the Fire, and of a Yellowifh colour:
and if You offer to put off to him a
piece of Brafs for a piece of Gold, he
will prefently refufe it, and (if he underftand Mettals) tell You, that though
Your Brafs be coloured like it, 'tis not
fo heavy, nor fo malleable, neither will
it like Gold refift the utmoft brunt of
the Fire, or refift *Aqua Fortis*: and if
You ask Men what they mean by a
Ruby, or Niter, or a Pearl, they will
ftill make You fuch Anfwers, that You
may clearly perceive, that whatever
Men talk in Theory of Subftantial
Forms, yet That, upon whofe account
they really diftinguifh any one Body
from

from others, and refer it to this or that *Species* of Bodies, is nothing but an Aggregate or Convention of such Accidents, as most men do by a kind of Agreement (for the Thing is more Arbitrary then we are aware of) think necessary or sufficient to make a Portion of the Universal Matter belong to this or that Determinate *Genus* or *Species* of Natural Bodies. And therefore not onely the Generality of Chymists, but diverse Philosophers, and, what is more, some Schoolmen themselves, maintain it to be possible to Transmute the ignobler Mettals into Gold; which argues, that if a Man could bring any Parcel of Matter to be Yellow, and Malleable, and Ponderous, and Fixt in the Fire, and upon the Test, and Indissoluble in *Aqua Fortis*, and in some to have a concurrence of all those Accidents, by which Men try True Gold from False, they would take it for True Gold without scruple. And in this case the generality

ty of Mankind would leave the School-
Doctors to difpute, whether being a
Factitious Body, (as made by the Chymifts art,) it have the Subftantial Form
of Gold, and would upon the account of
the Convention of the frefh!y mention'd Accidents let it pafs Current a-
mongft them, notwithftanding moft
Mens greater care, not to be deceived in
a matter of this nature then in any other.
And indeed, fince to every Determinate
Species of Bodies, there doth belong
more then One Quality, and for the moft
part a concurrence of Many is fo Effential to That fort of Bodies, that the
want of any of them is fufficient to exclude it from belonging to that *Species*:
there needs no more to difcriminate
fufficiently any One kind of Bodies
from all the Bodies in the World, that
are not of that kind; as the Chymifts
Luna fixa, which they tell us wants not
the Weight, the Malleableneffe, nor
the Fixtnefs, nor any other property of
Gold,

Gold, except the Yellowneffe, (which makes them call it White Gold,) would by reafon of that want of Colour be eafily known from true Gold. And you will not wonder at this, if you confider, that thofe Sphæres and Parallelopipedons differ but in Shape, yet this difference alone is the ground of fo many others, that *Euclid* and other Geometricians have demonftrated, I know not how many Properties of the one, which do no way belong to the other; and† *Ariftotle* himfelf fomewhere tels us, That a Sphære is compof'd of Brafs and Roundnefs. And I fuppofe it would be thought a Man's own fault, if he could not diftinguifh a Needle from a File, or a Key from a pair of Sciffors, though thefe being all made of Iron, and differing but in Bigneffe and Shape, are lefs remarkably diverfe then Natural Bodies, the moft part of which differ from each other in far more Accidents

---

† *Arift. Metapb. lib.7. cap. 8.*

then

then Two. Nor need we think that
Qualities being but Accidents, they
cannot be *essential* to a Natural Body;
for Accident, as I formerly noted ﹥ is
sometimes oppos'd to Substance, and
sometimes to Essence: and though an
Accident can be but accidental to Matter, as it is a Substantial thing, yet it
may be essential to this or that particular Body; as in *Aristotle's* newly mention'd Example, though Roundness is
but Accidental to Brass, yet 'tis Essential to a Brasen Sphære;because,though
the Brasse were devoid of Roundnesse,
(as if it were Cubical, or of any other
figure,) it would still be a Corporeal
Substance, yet without that Roundness
it could not be a Sphære: wherefore
since an Aggregate or Convention of
Qualities is enough to make the portion of Matter 'tis found in,what it is, and
denominate it of this or that Determinate sort of Bodies; and since those
Qualities, as we have seen already, do ·
themselves

themfelves proceed from thofe more
Primary and Catholick affections of
Matter, Bulk, Shape, Motion or Reft,
and the Texture thence refulting, why
may we not fay, that the Form of a Body being made up of thofe Qualities u-
nited in one Subject, doth likewife confift in fuch a Convention of thofe newly nam'd Mechanical Affections of
Matter, as is neceffary to conftitute a
Body of that Determinate kind. And fo,
though I fhall for brevities fake retain
the word *Forme*, yet I would be underftood to mean by it, not a Real *Sub-*
ftance diftinct from Matter, but onely
the Matter it felf of a Natural Body,
confider'd with its peculiar manner of
Exiftence, which I think may not in-
conveniently be call'd either its *Specifi-*
cal or its *Denominating State*, or its *Ef-*
fential Modification, or, if you would
have me exprefs it in one word , its
Stamp: for fuch a Convention of Accidents is fufficient to perform the Offi-
ces

ces that are neceſſarily requir'd in what
Men call a Forme, ſince it makes the
Body ſuch as it is, making it appertain
to this or that Determinate Species of
Bodies, and diſcriminating it from all
other Species of Bodies whatſoever: as
for Inſtance, Ponderouſneſs, Ductility,
Fixtneſſe, Yellowneſs, and ſome other
Qualities, concurring in a portion of
Matter, do with it conſtitute Gold, and
making it belong to that Species we
call Mettals, and to that ſort of Mettals
we call Gold, do both denominate and
diſcriminate it from Stones, Salts, Mar-
chaſites, and all other ſorts of Bodies
that are not Mettals, and from Silver,
Braſs, Copper, and all Mettals except
Gold. And whereas 'tis ſaid by ſome,
that the Forme alſo of a Body ought to
be the Principle of its Operations, we
ſhall hereafter conſider in what ſenſe
That is to be admitted or rejected; in
the mean time it may ſuffice us, that even
in the Vulgar Philoſophy 'tis acknow-
G ledg'd,

ledg'd, that Natural Things for the moſt
part operate by their Qualities, as Snow
dazles the Eyes by its Whiteneſs, and
Water ſcatter'd into drops of Rain falls
from the Clouds upon the account of
its Gravity. To which I ſhall adde, that
how great the power may be, which a
Body may exerciſe by virtue of a ſingle
Quality, may appear by the Various and
oftentimes Prodigious Effects, which
Fire produces by its Heat, when thereby it melts Mettals, calcines Stones,
deſtroves whole Woods and Cities &c.
And if ſeveral Active Qualities convene
in one Body, (as that which in our Hypotheſis is meant by Forme, uſually
compriſes ſeveral of them,) what great
things may be thereby perform'd, may
be ſomewhat gueſs'd at by the ſtrange
things we ſee done by ſome Engines,
which, being, as Engins, undoubtedly devoid of Subſtantial Forms, muſt do
thoſe ſtrange things they are admir'd
for, by virtue of thoſe Accidents, the
Shape

Shape, Size, Motion, and Contrivance, of their parts. Not to mention, that in our Hypothefis, befides thofe Operations that proceed from the Effential Modification of the Matter, as the Body (compof'd of Matter and neceffary Accidents) is confider'd *per modum unius*, as one Entire Corporeal Agent, it may in diverfe cafes have other Operations, upon the account of thofe particular Corpufcles, which though they concurre to compofe it, and are in reference to the whole confider'd but as its parts, may yet retain their own particular Nature, and diverfe of the peculiar Qualities: as in a Watch, befides thofe things which the Watch performs as fuch, the feveral parts whereof it confifts, as the Spring, the Wheels, the String, the Pins, &c. may have each of them its peculiar Bulk, Shape, and other Attributes, upon the account of one or more of which, the Wheel or Spring &c. may do other things then

what

what it doth, as meerly a Conftituent part of the Watch. And fo in the Milk of a Nurfe, that hath fome hours before taken a Potion, though the Corpufcles of the purging Medicine appear not to fenfe diftinct from the other parts of the Milk, which in far greater numbers concurre with them, to conftitute that white Liquor, yet thefe Purgative Particles, that feem but to be part of the Matter whereof the Milk confifts, do yet fo retain their own Nature and Qualities, that being fuck'd in with the reft by the Infant, they quickly difcriminate and difcover themfelves by purging him. But of this Subject more hereafter.

(Of Generation, Corruption, and Alteration.)

VIII. IT now remains that we declare, what, according to the Tenour of our Hypothefis, is to be meant by *Generation, Corruption,* and *Alteration*;

(Three

(Three Names, that have very much puzled and divided Philofophers.) In order hereunto we may confider,

1. That there are in the World great ftore of Particles of Matter, each of which is too fmall to be, whilft fingle, Senfible; and being Entire, or Undivided, muft needs both have its Determinate Shape, and be very Solid. Infomuch, that though it be *mentally*, and by Divine Omnipotence divifible, yet by reafon of its Smalnefs and Solidity, Nature doth fcarce ever actually divide it; and thefe may in this fenfe be call'd *Minima* or *Prima Naturalia*.

2. That there are alfo Multitudes of Corpufcles, which are made up of the Coalition of feveral of the former *Minima Naturalia*; and whofe Bulk is fo fmall, and their Adhæfion fo clofe and ftrict, that each of thefe little Primitive Concretions or Clufters (if I may fo call them) of Particles is fingly below the difcernment of Senfe, and though not

abfo-

absolutely indivisible by Nature into
the *Prima Naturalia* that compos'd it,
or perhaps into other little Fragments,
yet, for the reasons freshly intimated,
they very rarely happen to be actually
dissolv'd or broken, but remain entire
in great variety of sensible Bodies, and
under various forms or disguises. As,
not to repeat, what we lately mention'd,
of the undestroy'd purging Corpuscles
of Milk; we see, that even Grosser and
more compounded Corpuscles may
have such a permanent Texture: For
Quickfilver, for instance, may be turn'd
into a red Powder for a Fusible and
Malleable Body, or a Fugitive Smoak,
and disguis'd I know not how many o-
ther wayes, and yet remain true and re-
coverable Mercury. And these are as
it were the Seeds, or immediate Principles of many sorts of Natural Bodies,
as Earth, Water, Salt, &c. and those
singly insensible, become capable, when
united, to affect the Sense: as I have
try'd,

try'd, that if good Camphire be kept a
while in pure Spirit of Wine, it will
thereby be- reduc'd into fuch Little
parts, as totally to difappear in the Liquor, without making it look lefs clear
then fair Water, and yet, if into this
Mixture you pour a competent quantity of Water, in a moment the fcatter'd
Corpufcles of the Camphire will, by re-
uniting themfelves, become White,
and confequently Vifible , as before
their Difperfion.

3. That as well each of the *Minima
Naturalia*, as each of the Primary Clufters above mention'd, having its own
Determinate Bulk & Shape, when thefe
come to adhere to one another, it muft
alwaies happen, that the Size, and *often*,
that the Figure of the Corpufcle com-
poi'd by their Juxta-pofition and Cohæfion, will be chang'd: and *not feldome*
too, the Motion either of the one, or the
other, or both, will receive a new Tendency, or be alter'd as to its Velocity,

G 4   or

or otherwife. And the like will happen, when the Corpufcles, that compofe
a Clufter of Particles, are disjoyn'd, or
any thing of the little Mafs is broken
off. And whether any thing of Matter
be added to a Corpufcle, or taken from
it in either cafe, (as we juft now intimated,) the Size of it muft neceffarily be
alter'd, and for the moft part the Figure
will be fo too, whereby it will both acquire a Congruity to the Pores of fome
Bodies, (and perhaps fome of our Senfories,) and become Incongruous to
thofe of others, and confequently be
qualify'd, as I fhall more fully fhew you
hereafter, to operate on diverfe occafions, much otherwife then it was fitted
to do before.

4. That when many of thefe infenfible Corpufcles come to be affociated
into one vifible Body, if many or moft
of them be put into Motion, from what
caufe foever the Motion proceeds, That
it felf may produce great Changes, and
new

new Qualities in the Body they compofe; for *not onely* Motion may perform
much , even when it makes not any
vifible Alteration in it, as Air put into
fwift Motion, (as when it is blown out
of Bellows) acquires a new Name, and
is call'd *Wind*, and to the Touch appears
far colder then the fame *Air* not fo
form'd into a Stream: and Iron, by being
briskly rubb'd againft Wood or other
Iron, hath its fmall parts fo agitated, as
to appear hot to our Senfe: *but* this
Motion oftentimes makes vifible Alterations in the Texture of the Body into
which it is receiv'd , for alwaies the
Moved parts ftrive to communicate
their Motion, or fomewhat of the degree of it, to fome parts that were before either at Reft, or otherwife mov'd,
and oftentimes the fame Mov'd parts
do thereby either disjoyn, or break
fome of the Corpufcles they hit againft,
and thereby change their Bulk , or
Shape, or both, and either drive fome
of

of them quite out of the Body, and perhaps lodge themselves in their places,
or else associate them anew with others.
Whence it usually follows, that the
Texture, is for a while at least, and, unlesse it be very stable and permanent,
for good and all, very much alter'd, and
especially, in that the Pores or little In-
tervals intercepted betwixt the component Particles, will be chang'd as to
Bignefs, or Figure, or both, and so will
ceafe to be commenfurate to the Corpufcles that were fit for them before,
and become commenfurate to such Corpufcles of other Sizes and Shapes, as
till then were incongruous to them.
Thus we fee that Water, by loofing
the wonted agitation of its parts, may
acquire the Firmneffe and Brittleneffe
we find in Ice, and loofe much of the
Tranfparency it had whilft it was a Liquor. Thus alfo by very hard rubbing
two pieces of Refinous Wood againft
one another, we may make them throw

out

out diverse of their loofer parts into
Steams and vifible Smoak, and may, if
the Attrition be duely continued, make
that commotion of the parts fo change
the Texture of the whole, as afterwards
to turn the fuperficial parts into a kind
of Coal. And thus Milk, efpecially in
hot weather, will by the inteftine, though
languid, Motions of its parts, be in a
fhort time turn'd into a thinner fort of
liquor then Milk, and into Cream, and
this (laft nam'd) will by being barely
agitated in a Churn, be turn'd in a fhor-
ter time into that Unctuous and confiftent Body we call Butter, and into thin,
fluid, and fower Butter-milk. And thus
(to difpatch) by the bruifing of Fruit,
the Texture is commonly fo chang'd,
that as we fee particularly in Apples,
that the Bruif'd part foon comes to be
of another nature then the Sound part,
the one differing from the other both
in Colour, Taft, Smell, and Confiftence.
So that (as we have already inculcated)

Local

Local Motion hath, of all other affections of Matter, the greateſt Intereſt in
the Altering and Modifying of it, ſince
it is not onely the Grand *Agent* or *Efficient* among Second Cauſes, but is alſo
oftentimes one of the principal things
that *conſtitutes the Forme* of Bodies: as
when two Sticks are ſet on fire by long
and vehement Attrition, Local *Motion*
is not onely that which kindles the
Wood, and ſo as an Efficient produces
the Fire, but is That which principally
concurrs to give the produced Stream
of ſhining Matter, the name and nature
of Flame: and ſo it concurrs alſo to conſtitute all Fluid Bodies.

5. And that ſince we have formerly
ſeen, that 'tis from the Size, Shape, and
Motion of the ſmall parts of Matter,
and the Texture that reſults from the
manner of their being diſpoſ'd in any
one Body, that the Colour, Odour,
Taſt, and other qualities of that Body
are to be deriv'd, it will be eaſie for us
to

to recollect, That such Changes cannot
happen in a portion of Matter, without
so much varying the Nature of it, that
we need not deride the antient Atomifts, for attempting to deduce the *Ge-*
neration and *Corruption* of Bodies from
the fam'd σύγκρισις ϗ Διάκρισις, the *Conven-*
tion and *Diffolution*, and the *Alterations*
of them, from the *tranfpofition* of their
(*fuppof'd*) Atoms: For though indeed
Nature is wont in the Changes fhe
makes among things Corporeal, to imploy all the *three* wayes, as well in *Alte-*
rations, as *Generations* and *Corruptions*;
yet if they onely meant, as probably e-
nough they did, That of the *three* waies
propof'd, the Firft was wont to be the
Principal in the *Generation* of Bodies,
the fecond in the *Corruption*,& the third
in their *Alterations*, I fhall not much
oppofe this Doctrine: though I take
the Local Motion or *Tranfpofition* of
Parts,in the fame portion of Matter, to
bear a great ftroak as well in reference

to

to *Generation* and *Corruption*, as to *Alteration*: as we fee when Milk, or Flefh, or Fruit, without any remarkable addition or lofs of parts turns into Maggots, or other Infects; and as we may more confpicuoufly obferve in the Præcipitation of Mercury without addition, in the Vitrification of Mettals, and other Chymical Experiments to be hereafter mention'd.

Thefe things premif'd, it will not now be difficult to comprife in few words fuch a Doctrine, touching the *Generation*, *Corruption*, and *Alteration* of Bodies, as is fuitable to our *Hypothefis*, and the former Difcourfe. For if in a parcel of Matter there happen to be produc'd (it imports not much how) a Concurrence of all thofe Accidents, (whether thofe onely, or more) that Men by tacite agreement have thought *neceffary* and *fufficient* to conftitute any one Determinate *Species* of things corporeal, then we fay, That a Body belonging

longing to that *Species*, as fuppofe a
Stone, or a Mettal, is *Generated*, or produc'd *de novo*. Not that there is really
any thing of *Subftantial* produc'd, but
that thofe parts of Matter, that did indeed before præexift, but were either
fcatter'd and fhar'd among other Bodies,
or at leaft otherwife difpof'd of, are
now brought together, and difpof'd of
after the manner requifite, to entitle
the Body that refults from them to a
new Denomination, and make it appertain to fuch a Determinate *Species* of
Natural Bodies, fo that no new *Sub-*
ftance is in Generation *produc'd*, but
onely That, which was *præexiftent*, ob-
teins a new *Modification*, or manner
of Exiftence. Thus when the Spring,
and Wheels, and String, and Balance,
and Index &c. neceffary to a Watch,
which lay before fcatter'd, fome in one
part, fome in another of the Artificer's
Shop, are firft fet together in the Order
requifite to make fuch an Engine, to
shew

shew how the time passes, a watch is
said to be *made*: not that any of the
mention'd Material parts is *produc'd de
novo*, but that till then the divided Matter was not so *contriv'd* and put together, as was requisite to constitute such
a thing, as we call a Watch. And so
when Sand and Ashes are well melted
together, and suffer'd to cool, there is
Generated by the Colliquation that
sort of Concretion we call *Glaß*, though
it be evident, that its Ingredients were
both præexistent, and do but by their
Association obtain a New manner of existing together. And so when by the
Churning of Creame, Butter and Butter-milk are generated, we find not any
thing Substantial Produc'd *de novo* in
either of them, but onely that the *Se-
rum*, and the fat Corpuscles, being put
into Local Motion, do by their frequent
Occursions extricate themselves from
each other, and associate themselves in
the new manner, requisite to constitute
                                      the

the Bodies, whose names are given them.

And as a Body is said to be *generated*, when it first appears clothed with all those Qualities, upon whose Account Men have been pleas'd to call some Bodies *Stones*; others, *Mettals*; others, *Salts*, &c. so when a Body comes to loose *all* or *any* of those Accidents that are *Essential*, and necessary to the constituting of such a Body, it is then said to be *corrupted* or destroy'd, and is no more a Body of *that Kind*, but looses its Title to its former Denomination. Not that any thing *Corporeal* or Substantial *perishes* in this *Change*, but onely that the Essential Modification of the Matter is destroy'd: and though the Body be still a *Body*, (no Natural Agent being able to *annihilate* Matter,) yet 'tis no longer *such a Body*, as 'twas before, but perisheth in the capacity of a Body of that Kind. Thus if a Stone, falling upon a Watch, break it to pieces;

H   *as,*

as, when the Watch was made there was no new Subſtance produc'd, all the Material parts (as the Steel , Braſs, String, &c.) being præexiſtent ſome where or other, (as in Iron,and Copper-Mines, in the Bellies of thoſe Animals, of whoſe Guts Men uſe to make Strings;) *ſo* not the leaſt part of the Subſtance of the Watch is loſt , but onely diſplac'd and ſcatter'd; and yet that Portion of Matter ceaſes to be a *VVatch* as it was before. And ſo (to reſume our late Example)when Creame is by Churning turn'd into Butter, and a Serous Liquor, the parts of the Milk remain aſſociated intothoſe two Bodies, but the White Liquor periſheth in the capacity of Milk. And ſo when Ice comes to be thaw'd in exactly cloſ'd Veſſels,though the Corruption be pro-duc'd onely (for ought appears) by in-troducing a new Motion and Diſpoſiti-on into the parts of the Frozen Water, yet it thereupon ceaſes to be *Ice*, howe-

ver

ver it be as much *VVater*, and confequently as much a *Body*, as before it was
frozen or thaw'd. Thefe and the like
Examples may teach us rightly to underftand that common Axiom of Naturalifts, *Corruptio unius eft generatio
alterius*; *& è contrà:* for fince it is acknowledged on all hands, that Matter
cannot be annihilated, and fince it appears by what we have faid above, that
there are fome Properties, namely *Size,
Shape, Motion,* (or in its abfence, *Reft,)*
that are infeparable from the actual parts
of Matter; and fince alfo the Coalition
of any competent number of thefe parts
is fufficient to conftitute a Natural Body, endow'd with diverfe fenfible Qualities; it can *fcarce* be otherwife, but that
the fame Agents, that fhatter the
Frame, or *deftroy* the Texture of one
Body, will by fhuffling them together,
and difpofing them after a New manner,
bring them to *conftitute* fome new fort
of Bodies: As the fame thing, that by

burning deftroyes Wood, turns it into
Flame, Soot, and Afhes. Onely I doubt,
whether the Axiome do generally hold
true, if it be meant, That *every Corrup-*
tion muft end in the Generation of a Body,
belonging to fome particular Species of
things, unleffe we take Powders and
fluid Bodies indefinitely for *Species* of
Natural Bodies; fince it is plain, there
are multitudes of Vegetables, and other
Concretions, which, when they rot, do
not, as fome others do, turn into
Worms, but either into fome flimy or
watery Subftance, or elfe (which is the
moft ufuall) they crumble into a kind
of Duft or Powder, which, though
look'd upon as being the Earth, into
which rotten Bodies are at length refolv'd, is very far from being of an Elementary nature, but as yet a Compounded Body, retaining fome, if not many,
Qualities, which often makes the Duft
of one fort of Plant or Animal differ
much from that of another. And This
will

will fupply me with this Argument *Ad hominem*, viz. That fince inthofe *violent Corruptions* of Bodies, that are made by Outward Agents, fhattering them into pieces, if the Axiome hold true, the New *Bodies emergent* upon the Diffolution of the Former, muft be really *Natural Bodies*, as (indeed divers of the Moderns hold them to be,) and Generated according to the courfe of Nature; as when Wood is deftroy'd by Fire, and turn'd partly into Flame, partly into Soot, partly into Coals, and partly into Afhes; I hope we may be allow'd to conclude, That thofe *Chymical Productions*, which fo many would have to be but *Factitious Bodies*, are *Natural ones*, and regularly Generated. For it being the fame Agent, the Fire, that operates upon Bodies, whether they be expof'd to it in clofe Glaffes, or in Chimnies, I fee no fufficient reafon, why the Chymical Oyls, and Volatile Salts, and other things which

H 3 · Spa-

Spagirites obtain from mixt Bodies, should not be accounted Natural Bodies, as well as the Soot, and Ashes, and Charcoal, that by the same fire are obtain'd from Kindled Wood.

But before we passe away from the mention of the Corruption of Bodies, I must take some notice of what is call'd their *Putrefaction*. This is but a Peculiar kind of Corruption, wrought slowly (whereby it may be distinguish'd from Destruction by Fire, and other nimble Agents) in Bodies: it happens to them for the most part by means of the Air, or some other Ambient Fluid, which by penetrating into the Pores of the Body, and by its agitation in them, doth usually call out some of the more Agile and lesse entangled parts of the Body, and doth almost ever loosen and dislocate the parts in general, and thereby so change the Texture, and perhaps too the Figure, of the Corpuscles, that compose it, that the Body, thus chang'd,

acquires

acquires Qualities unfuitable to its Former Nature, and for the moft part offenfive to Our Senfes , efpecially of
Smelling and Tafting: which laft claufe
I therefore adde, not onely becaufe the
Vulgar look not upon the Change of
an Egge into a Chick as a *Corruption*,
but as a *Perfection* of the Egge; but becaufe alfo I think it not improbable,
that if by fuch flow Changes of Bodies,
as make them loofe their former Nature, and might otherwife paffe for *Pu-*
trefaction, many Bodies fhould acquire
better Sents or Tafts then before; or if
Nature, Cuftom, or any other caufe
fhould much alter the Texture of our
Organs of Tafting and Smelling, it
would not perhaps be fo well agreed on
what fhould be call'd Putrefaction, as
that imports an *impairing Alteration*,
but Men would find fome favourabler
Notion for fuch Changes.   For I obferve, that Medlars, though they acquire
in length of time fuch a Colour and
Softnefs

Softnefs as rotten Apples, and other pu-
trify'd Fruits do, yet, becaufe their Taft
is not then harfh as before, we call that
Ripenefs in them, which otherwife we
fhould call *Rottenneſs*. And though upon the Death of a fourfooted Beaft, we
generally call that Change, which happens to the Flefh or Bloud, Putrefaction, yet we paffe a more favourable judgment upon That, which happens to the
Flefh and other fofter parts of that A-
nimal, (whether it be a kind of large
Rabbets, or very fmall and hornleffe
Deer,) of which in *China*, and in the
Levant they make Mufk; becaufe by
the Change, that enfues the Animals
death, the Flefh acquires not an *odious*,
but a *grateful Smell*. And we fee, that
fome Men, whofe Appetites are grati-
fied by Rotten Cheefe, think it Then
not to have *degenerated*, but to have attain'd its *beft State*, when having loft its
former Colour, Smell, and Taft, and,
which is more, being in great part
turn'd

turn'd into thofe Infe&s call'd Mites, 'tis both in a Philofophical fenfe *cor-rupted*, and in the æftimate of the generality of Men grown *Putrid*. But becaufe it very feldom happens, that a Body by Generation acquires no other Qualities, then juft thofe that are abfolutely *neceffary*, to make it belong to the *Species* that Denominates it; therefore in moft Bodies there are diverfe other Qualities that may *be* there, or may be *miffing*, without Effentially changing the Subje&: as Water may be clear or muddy, odorous or ftinking, and ftill remain Water; and Butter may be white or yellow, fweet or rancid, confiftent or melted, and ftill be call'd Butter. ·Now therefore whenfoever a Parcel of Matter does *acquire* or *loofe* a Quality, that is not *Effential* to it, That Acquifition or Loffe is diftin&tly call'd *Alteration*, (or by fome, *Mutation*:) the Acquift onely of the Qualities that are abfolutely *neceffary* to conftitute its Ef-

fential

fential and Specifical difference, or the Lofs of any of *thofe* Qualities, being fuch a Change as mult not be call'd meer *Alteration*, but have the particular name of Generation or Corruption; both which according to this Doctrine appear to be but feveral *Kinds of Alteration*, taken in a large fenfe, though they are diftinguifh'd from it in a more ftrict and Limited acception of that Terme.

And here we have a fair Occafion to take notice of the Fruitfulneffe and Extent of our Mechanical Hypothefis: For fince according to our Doctrine, the World we live in is not a Moveleffe or Indigefted Mafs of Matter, but an ᾽Αυτόματον, or *Self-moving Engine*, wherein the greateft part of the common Matter of all Bodies is alwaies (though not ftill the fame parts of it) in Motion, & wherein Bodies are fo clofe fet by one another, that (unleffe in fome very few and extraordinary, and as it were Præternatural

ternatural cafes) they have either no
Vacuities betwixt them, or onely here
and there interpos'd, and very fmall
ones. Aud fince, according to us, the
various *manner* of the *Coalition* of feveral *Corpufcles* into one vifible *Body* is e-
nough to give them a peculiar Texture,
and thereby fitt them to exhibit divers
fenfible Qualities, and to become a Body, fometimes of one Denomination,
and fometimes of another; it will very
naturally follow, that from the various
Occurfions of thofe innumerable fwarms
of little Bodies, that are mov'd to and
fro in the World, there will be many
fitted to ftick to one another, and fo
compofe Concretions ; and many
(though not in the felf fame place) dif-
joyn'd from one another, and agitated
apart; and multitudes alfo that will be
driven to affociate themfelves, now with
one Body, and prefently with another,
And if we alfo confider on the one fide,
that the Sizes of the fmall Particles of
Matter

Matter may be very *various*, their Figures almoſt *innumerable*, and that if a
parcel of Matter do but happen to ſtick
to one Body, it may chance to give it a
new Qualitv, and if it adhere to another,
or hit againſt ſome of its Parts, it may
conſtitute a Body of another Kind; or if
a parcel of Matter be knockt off from
another, it may barely by That, leave
It, and become it ſelf of another Nature
then before. If, I ſay, we conſider
theſe things on the one ſide; and on the
other ſide, that (to uſe *Lucretius* his
Compariſon) all that innumerable multitude of Words, that are contain'd in
all the Languages of the World, are
made of the various Combinations of
ſome of the 24 Letters of the Alpha-
bet; 'twill not be hard to conceive, that
there may be an incomprehenſible variety of Aſſociations and Textures of
the Minute parts of Bodies, and conſequently a vaſt Multitude of Portions of
Matter endow'd with ſtore enough of
differing

differing Qualities, to deferve diftinct
Appellations, though for want of heed-
fulneffe and fit Words, Men have not
yet taken fo much notice of their leffe
obvious Varieties, as to fort them as
they deferve, and give them diftinct and
proper Names. So that though I
would not fay, that Any thing can immediately be made of Every thing, as a
Gold Ring of a *VVedge* of Gold, or
Oyl, or Fire of Water; yet fince Bodies,
having but one common Matter, can be
differenc'd but by Accidents, which
feem all of them to be the Effects and
Confequents of Local Motion, I fee
not, why it fhould be abfurd to think,
that (at leaft among Inanimate Bodies)
by the Intervention of fome very fmall
Addition or *Subftraction* of Matter,
(which yet in moft cafes will fcarce be
needed,) and of an orderly *Series of Alterations*, difpofing by degrees the Matter to be tranfmuted, almoft of any
thing, may at length be made Any
thing:

thing: as, though out of a *wedge* of Gold one cannot immediately make a *Ring*, yet by either Wyre-drawing that Wedge by degrees, or by melting it, and casting a little of it into a Mould, That thing may easily be effected. And so though Water cannot immediately be transmuted into Oyl, and much less into Fire, yet if you nourish certain Plants with Water alone, (as I have done,) 'till they have assimilated a great quantity of Water into their own Nature, You may, by committing this Transmuted Water (which you may distinguish and separate from that part of the Vegetable you first put in) to Distillation in convenient Glasses, obtain, besides other things, a true Oyl, and a black combustible Coal, (and consequently Fire,) both of which may be so copious, as to leave no just cause to suspect, that they could be any thing neer afforded by any little Spirituous parts, which may be præsum'd to have

been

been communicated by that part of the Vegetable, that is firſt put into the water, to that far greater part of it, which was committed to Diſtillation.

But, *Pyrophilus*, I perceive the Difficulty and Fruitfulneſſe of my Subject, have made me ſo much more prolix then I intended, that it will not now be amiſs to Contract the Summary of our *Hypotheſis*, and give you the Main Points of it with little or no Illuſtration, and without particular Proofs in a few words. We teach then (but without peremptorily aſſerting it,)

Firſt, That the Matter of all Natural Bodies is the Same, namely a Subſtance extended and impenetrable.

2. That all Bodies thus agreeing in the ſame common Matter, their Diſtinction is to be taken from thoſe Accidents that do diverſify it.

3. That Motion, not belonging to the Eſſence of Matter,(which retains its whole Nature, when 'tis at Reſt,) and

not

not being Originally producible by o-ther Accidents, as They are from It, may be look'd upon as the Firſt and chief *Mood* or Affection of Matter.

4. That Motion, variouſly deter-min'd, doth naturally divide the Matter it belongs to, into actual Fragments or Parts; and this Diviſion obvious Ex-perience, (and more eminently, Chy-mical Operations)manifeſt to have been made into parts exceedingly *minute*, and very often, too minute to be ſingly perceiveable by our Senſes.

5. Whence it muſt neceſſarily fol-low, that each of theſe Minute Parts, or *minima Naturalia* (as well as every particular Body, made up by the Coali-tion of any number of them,) muſt have its Determinate *Bigneſſe* or *Size*, and its own *Shape*. And theſe three, namely *Bulk*, *Figure*, and either *Motion* or *Reſt*, (there being no Mean between theſe two) are the three *Primary* and moſt *Catholick Moods* or Affections of •

the

the *infenfible* parts of Matter, confider'd *each* of them *apart*.

6. That when *diverfe* of them are confider'd *together*, there will neceffarily follow here Below both a certain *Pofition* or *Pofture* in reference to the Horizon (as Erected, Inclining, or Level) of each of them, and a certain *Order*, or placing before, or behind, or befides one another; (as when in a company of Souldiers, one ftands *upright*, the other *ftoops*, the other *lyes along* upon the Ground, they have various *Poftures*; and their being plac'd *befides* one another in Ranks, and *behind* one another in Files, are Varieties of their *Order*:) and when many of thefe fmall parts are brought to Convene into one Body from their *primary Affections*, and their Difpofition, or *Contrivance* as to *Pofture* and *Order*, there refults That, which by one Comprehenfive Name we call the *Texture* of that Body. And indeed thefe feveral Kinds of *Location*,

I   (to

to borrow a Scholaſtical Terme,) attributed (in this 6ᵗʰ number) to the Minute Particles of Bodies, are ſo neer of
Kinne, that they ſeem all of them refer-
rable to (that One Event of their Convening,) *Scituation*, or *Poſition*. And
theſe are the Affections that belong to
a Body, as it is conſider'd in it ſelf, without relation to *ſenſitive* Beings, or to
other Natural Bodies.

7. That yet, there being Men in
the World, whoſe Organs of Senſe are
contriv'd in ſuch differing wayes, that
one Senſory is fitted to receive Impreſſions from ſome, and another from o-
ther ſorts of External Objects, or Bodies without them, (whether theſe act
as Entire Bodies, or by *Emiſſion* of their
Corpuſcles, or by *propagating* ſome
Motion to the Senſory,) the Perceptions of theſe Impreſſions are by men
call'd by ſeveral Names, as *Heat*, *Colour*,
Sound, *Odour*; and are commonly imagin'd to proceed from certain diſtinct
and

and peculiar Qualities in the External Object, which have some resemblance to the Ideas, their action upon the Senses excites in the Mind; though indeed all these Sensible Qualities, and the rest that are to be met with in the Bodies without us, are but the Effects or Consequents of the above mentioned *primary Affections* of Matter, whose Operations are diversify'd according to the nature of the Sensories, or other Bodies they work upon.

8. That when a Portion of Matter, either by the *accession* or *Recesse* of Corpuscles, or by the *transposition* of those it consisted of before, or by any *two* or *all* of these waies, happens to obtain a *concurrence of all* those Qualities, which Men commonly agree to be *necessary* and *sufficient* to Denominate the Body, which hath them, either a *Mettal*, or a *Stone*, or the like, and to rank it in any peculiar and determinate *Species* of Bodies, Then a Body of that Denomina-

tion

tion is faid to be *Generated.*

9. This *Convention of Effential Accidents* being taken (not any of them Apart, but all) *together* for the Specifical Difference that *conftitutes* the Body, and *difcriminates* it from all other forts of Bodies, is by one Name, becaufe confider'd as one *collective* Thing, call'd its *Forme*, (as Beauty, which is made up both of Symmetry of Parts, and Agreeableneffe of Colours,) which is confequently but a certain *Character*, (as I fometimes call it,) or a *peculiar ftate of Matter*, or, if I may fo name it, an *Effential Modification*: a *Modification*, becaufe 'tis indeed but a Determinate *manner of Exiftence* of the Matter, and yet an *Effential Modification*, becaufe that though the concurrent Qualities be but Accidental to Matter, (which with others in ftead of Them, would be Matter ftill,) yet they are *effentially neceffary* to the Particular *Body*, which without thofe Accidents would

not

not be a Body of that Denomination, as a *Mettal* or a *Stone*, but of some other.

10. Now a Body being capable of many *other* Qualities, *besides* those, whose Convention is *necessary* to make up its Form; the *acquisition* or *lesse* of any such Quality is, by Naturalists in the more strict sense of that Terme, nam'd *Alteration*: as when Oyl comes to be frozen, or to change colour, or to grow rancid; but if all, or any of the Qualities, that are reputed *essential* to such a Body, come to be *lost* or *destroy'd*, that notable Change is call'd *Corruption*; as when Oyl being boyl'd takes fire, the Oyl is not said to be *alter'd* in the former sense, but *corrupted* or *destroy'd*, and the emergent Fire *generated*; and when it so happens, that the Body is *slowly corrupted*, and thereby also acquires *Qualities offensive to our Senses*, especially of *Smell* and *Tast*, (as when Flesh, or Fruit grows rotten,)

I 3   that

that kind of *Corruption* is by a more particular Name call'd *Putrefaction*. But neither in this, nor in any other kind of Corruption is there any thing *substantial* deftroy'd, (no fuch thing having been produc'd in Generation, and Matter it felf being on all hands acknowledged *incorruptible*,) but onely that *special connexion of the Parts*, or *manner of their Coexistence*, upon whofe account the Matter, whilft it was in its former ftate, was, and was call'd a Stone, or a Mettal, or did belong to any other Determinate *Species* of Bodies.

CON-

CONSIDERATIONS
and
EXPERIMENTS,
Touching the *Origine* of
QUALITIES and *FORMS*.

THE HISTORICAL PART.

The I. SECTION.

The I. Section,

Containing the *Observations*.

IN the foregoing Notes I have endeavoured with as much Clearnefs, as
the Difficulty of the Subject, and the
Brevity I was confined to, permitted to
give a Scheme or Summary of the Principles of the Corpufcularian Philofophy, as I apprehended them, by way of
a fhort Introduction to it, at leaft as far
as I judged neceffary for the better un-
derftanding of what is contain'd in our
Notes and Experiments concerning the
Productions and Changes of particular
Qualities. But though, I hope, I have
not fo affected Brevity, as to fall into
Obfcurity; yet fince thefe Principles
are built upon the *Phænomena* of Nature, and devif'd in order to the Explication

cation of them, I know not what I can do more proper to recommend them, then to fubjoyn fome fuch Natural *Phænomena*, as either induce me to take up fuch Notions, or which I was directed to find out by the Notions I had imbrac'd. And fince I appeale to the Teftimony of Nature to verifie the Doctrine I have been propofing, about the Origine and Production of Qualities, (for that of Formes will require a diftinct Difcourfe,) I think it very proper to fet down fome Obfervations of what Nature does, without being over-rul'd by the Power and Skill of Man, as well as fome Experiments wherein Nature is guided, and as it were Mafter'd by Art, that fo fhe may be made to atteft the Truth of our Doctrine, as well, when fhe difclofes her Self freely, and, if I may fo fpeak, of her Own accord, as when fhe is as it were Cited to make her Depofitions by the Induftry of Man. The Obfervations will be but

the

the more fuitable to our Defign for being Common and Familiar, as to the
Phænomena, though perhaps New e-
nough as to the Application to our
Purpofe. And as for the Experiments, becaufe thofe that belong more
immediately to this or that particular
Quality, may be met with in the Notes
that treat of It, I thought it not amiffe
that the Experiments fhould be both
Few in number, and yet fo Pregnant,
that every one of them fhould afford
fuch differing *Phænomena,* as may make
it applicable to more then One Quality.

I.

The Obfervation I will begin with
fhall be fetch'd from what happens in
the Hatching of an Egge. For as familiar and obvious a thing as it is, (efpecially after what the Learned *Fabricius ab Aqua pendente*, and a recenter A-
natomift have delivered about them,)
that there is a great Change made in
the

the substance of the Egge, when 'tis by
Incubation turn'd into a Chick: yet, as
far as i know, this Change hath not
been taken notice of, for the same purpose, to which I am about to apply it.

I consider then, that in a Prolifick
Egge, (for Instance that of a Hen,) as
well the Liquor of the Yolk, as that of
the White, is a Substance, as to sense,
Similar. For upon the same account
that Anatomists and Physicians call several parts of the humane Body, as
Bones, Membranes, &c. Similar, that is,
such, as that every Sensible part of it
hath the same Nature or Denomination with the whole, as every Splinter of
Bone is Bone, as every Shred of Skin
is Skin.

And though I find by distilling the
Yolks and Whites, they seem to be
Dissimilar Bodies, in regard that the
White of an Egge (for Example) will
afford Substances of a very differing
Nature, as Flegme, Salt, Oyl, and Earth,
yet

yet (not now to examine whether, or
how far thefe may be efteem'd Productions of the Fire, that are rather obtain'd from the White of the Egge,
then were præexiftent in it; not to mention this I fay,) it doth not appear by
Diftillation, that the White of an Egg,
is other then a Similar Body in the fenfe
above deliver'd. For it would be hard
to prove, that one part of the White
of an Egg will not be made to yield the
fame differing Subftances by Diftillation, that any other part does; and Bones
themfelves, and other hard parts of a
humane Body, that are confeffedly Similar, may by Diftillation be made to
afford Salt, and Phlegme, and Spirit,
and Oyl, and Earth, as well as the
White of an Egg.

This being thus fetled in the Firft
place, we may in the Next confider,
that by beating the White of an Egge
well with a Whisk, you may reduce it
from a fomewhat Tenacious into a Fluid
Body,

Body, though this Production of a Liquor be, as we elfewhere noted, effected
by a Divulfion, Agitation &c. of the
parts, that is in a word, by a Mechanical
change of the Texture of the Body.

In the Third place I confider, that
according to the exacteft Obfervations
of Modern Anatomifts, which our own
Obfervations do not contradict, the
Rudiments of the Chick, lodg'd in the
Cicatricula, or white Speck upon the
Coat of the Yolk, is nourifh'd, 'till it
have obtain'd to be a great Chick, onely by the White of the Egg; the Yolk
being by the Providence of Nature referv'd as a more ftrong and folid Aliment, till the Chick have abfum'd the
White, and be thereby grown great and
ftrong enough to digeft the Yolk; and
in effect you may fee the Chick furnifh'd not onely with all the neceffary,
but divers other parts, as Head, Wings,
Legs, and Beak, and Claws, whilft the
Yolk feems yet as it were untouch'd.
But

But whether this Obfervation about
the Entirenefs of the Yolk be precifely
true, is not much material to our prefent purpofe, nor would I be thought to
build much upon it; fince the Yolk it
felf, efpecially at that time, is wont to
be fluid enough, and to be a Liquor perhaps no lefs fo then the White was, and
That is enough for my prefent purpofe.

For in the Laft place I confider, that
the Nutritive Liquor of an Egg, which
is in it felf a Body fo very foft, that by
a little Agitation it may be made Fluid,
and is readily enough diffolvable in
common cold water, this very Subftance, I fay, being brooded on by the
Hen, will within two or three weeks be
tranfmuted into a Chick, furnifh'd with
Organical parts, as Eyes, Ears, Wings,
Legs, &c. of a very differing Fabrick,
and with a good number of Similar
ones, as Bones, Cartilages, Ligaments,
Tendons, Membranes, &c. which differ very much in Texture from one a-
nother;

nother; befides the Liquors, as Blood, Chyle, Gall, &c. contain'd in the folid parts: So that here we have out of the White of an Egg , which is a Subſtance Similar, Infipid, Soft, (not to call it Fluid,) Diaphanous, Colourleſſe, and readily diſſoluble in cold water, out of this Subſtance I fay; we have by the new and various Contrivement of the ſmall parts it confiſted of, an Animal, ſome of whoſe parts are not Tranſparent but Opacous; ſome of them Red, as the Bloud; ſome Yellow or Greeniſh, as the Gall; ſome White, as the Brain; ſome Fluid, as the Bloud, and other Juices; ſome. Confiſteut, as the Bones, Fleſh, and other. ſtable. parts of the Body; ſome Solid and Frangible, as the Bones, others Tough and Flexible, as the Ligaments, others Soft and looſly Cohærent , as the Marrow ; ſome without Springs, as many of the parts; ſome with Springs, as the Feathers; ſome apt to mingle readily with cold
water,

water, as the Bloud, the Gall; some not to be *so* dissolv'd in it, as the Bones, the Claws, and the Feathers; some well tasted, as the Flesh and Bloud; some very ill tasted, as the Gall, (for That I have purposely and particularly observ'd.) In a word, we have here produc'd out of such an uniforme Matter as the White of an Egg,

First, new kind of Qualities, as (besides Opacity) Colours, (whereof a single Feather will sometimes afford us Variety,) Odours, Tasts, and Heat in the Heart and Bloud of the Chick; Hardness, Smoothness, Roughness,&c.

Secondly, diverse other Qualities, that are wont to be distinguish'd from Sensible ones, as Fluidity (in the Bloud and aqueous humor of the Eye,)Consistency in the Grisles, Flesh, &c. Hardnesse, Flexibility, Springynesse, Toughness, unfitnesse to be dissolv'd in cold water, and several others. To which may probably be added

K   Thirdly,

Thirdly, fome Occult Properties as Phyficians obferve, that fome Birds, as young Swallows, young Magpies afford Specifick, or at leaft Noble Medicines, in the Falling ficknefs, Hyfterical Fits, and divers other Diftempers.

Fourthly, I very well forefee it may be objected, that the Chick with all its parts is not a Mechanically contriv'd Engine, but fafhion'd out of Matter by the Soul of the Bird, lodg'd chiefly in the *Cicatricula*, which by its Plaftick power fafhions the obfequious Matter, and becomes the Architect of its own Manfion. But not here to examine, whether any Animal, except Man, be other then a Curious Engine, I anfwer, that this Objection invalidates not what I intend to prove from the alledg'd Example. For let the Plaftick Principle be what it will, yet ftill, being a Phyfical Agent, it muft act after a Phyfical manner, and having no other Matter to work upon but the White of the Egg, it

it can work upon that Matter but as
Phyſical Agents, and conſequently can
but divide the Matter into minute parts
of ſeveral Sizes and Shapes, and by
Local Motion variouſly context them,
according to the Exigency of the Animal to be produc'd, though from ſo many various Textures of the produc'd
parts there muſt naturally emerge ſuch
differences of Colours, Taſts, and Con-
ſiſtencies, and other Qualities as we
have been taking notice of. That which
we are here to conſider, is not what is
the Agent or Efficient in theſe Productions, but what is done to the Matter to
effect them. And though ſome Birds
by an inbred Skill do very Artificially
build their Curious Neſts, yet cannot
Nature, that teaches them, enable them
to do any more then ſelect the Materials of their Neſts, and by Local Motion divide, tranſport, and connect them
after a Certain manner. And when Man
himſelf, who is undoubtedly an Intel-

ligent

ligent Agent, is to frame a Building or an Engine, he may indeed by the help of Reaſon and Art, contrive his Materials curiouſly and skilfully, but ſtill all he can do, is but to move, divide, tranſpoſe, and context the ſeveral parts, into which he is able to reduce the Matter aſſign'd him.

Nor need we imagine, that the Soul of that Hen, which having firſt produc'd the Egg, does after a while ſit on it, hath any peculiar Efficiency in hatching of a Chick: for the Egg will be well hatch'd by another Hen, though That which laid it be dead; and, which is more, we are aſſur'd by the Teſtimony of very good Authors, as well as of recent Travellers, that in ſome places, eſpecially in *Ægypt*, there needs no Bird at all to the Production of a Chick out of an Egg, ſince they hatch multitudes of Eggs by the regulated heat of Ovens, or Dunghils. And indeed, that there is a Motion or Agitation of the parts

parts of the Egg by the external heat, whereby it is hatch'd, is evident of its self, and not (as far as I know) deny'd by any, and that also the white Substance is abfumed and contexted, or contriv'd into the Body of the Chick, and its several parts, is manifeft to fenfe; especially if one hath the Curiofity to observe the progrefs of the Chicks Formation and Increment. But as 'tis evident, that as thefe two things, the Subftance of the White, and the Local Motion, wherein the External Heat neceffary to Incubation puts its parts, do eminently concurr to the Production of the Chick; fo that the Formative Power (whatever that be) doth any more then guide thefe Motions, and thereby affociate the fitted Particles of Matter after the manner requifite to conftitute a Chick, is that which I think will not eafily be evinc'd. And I might to what I faid of the Egg, adde feveral things touching the Generation of Vi-

K 3   viparous

viparous Animals, which the Learned
Fabricius ab Aqua pendente, as well as
some of the Antient Philosophers,
would have to be generated from an
Imperfect kind of Eggs: but I take the
Eggs of Birds to be much fitter to instance in, because they are things that
we have more at command, and wherewith we can conveniently make more
Trials and Observations; and especially
because in perfect Eggs the Matter to
be transmuted is more closely lock'd
up, and being kept from any visible
supply of Matter , confin'd to be
wrought upon by the External Heat,
and by its own Vital Principle within.

II.

Water being generally esteem'd an
Elementary Body, and being at least
far more Homogeneous then Bodies
here below are wont to be; it may make
very much for our present purpose to
shew, that Water it self, that is Fluid,
Tastlesse,

Taftlefs, Inodorous, Diaphanous, Colourlefs, Volatile, &c. may, by a differing Texture of its Parts, be brought to conftitute Bodies of Attributes very diftant from thefe. This I thought might be done, by nourifhing Vegetables with fimple water. For in cafe I could do fo, all, or the greateft part of that which would accrue to the Vegetable thus nourifh'd, would appear to have been materially but Water, with what Exotick Quality foever it may afterwards, when tranfmuted, be endow'd.

The Ingenious *Helmont* indeed mentions an Experiment fomewhat of this nature, though not to the fame purpofe, which he made by planting a Branch of Willow into a Pot full of Earth, and obferving the increafe of Weight he obtain'd after divers years, though he fed the Plant but with Rain water. And fome Learned Modern Naturalifts have conjectur'd at the eafy Tranfmutablenefle of Water, by what happens in

Gardens

Gardens and Orchards, where the fame
Showers or Rain after a long Drought
makes a great number of differing
Plants to flourifh. But though thefe
things be worthy of their Authors, yet
I thought they would not be fo fit for
my purpofe, becaufe it may be fpeci.
oufly enough objected, That the Rain
water does not make thefe Plants thrive
and flourifh, by immediately affording
them the Aliments they affimilate into
their own Subftance, but by proving a
Vehicle, that diffolves the Saline, and o-
ther Alimental Subftances of the Earth,
and dilutes both them and the nutritive Juice, which, in a part of the Plant
its felf, it may find too much thickned
by the Drought or Heat of the ambient
Air, and by this means it contributes
to the nourifhment of the Plant, though
it felf be infenfibly afterwards exhal'd
into vapours. And indeed Experience
fhews us, that feveral Plants, that thrive
not well without Rain water, are not
yet

yet nourifh'd by it alone, fince when
Corn in the Field, and Fruit-trees in
Orchards have confum'd the Saline and
Sulphureous Juices of the Earth, they
will not profper there, how much Rain
foever falls upon the Land, till the
Ground by Dung or otherwife be fupply'd again with fuch affimilable Juices.
Wherefore I rather chofe to attempt
the making of Plants grow in Viols
fill'd with Water, not onely to prevent
the forementioned Objection, and alfo
to make the Experiment leffe tedious,
but that I might have the pleafure of
feeing the progrefs of Nature in the
Tranfmutation of Water; and my Obfervations of this kind as Novelties, unmention'd by any other Writer, I fhew'd
divers Ingenious Freinds, who having
better Opportunities then I of ftaying
in one place, have attempted the like,
and made fuccefful Trials, which, I fuppofe, will not be conceal'd from the
publick. Of my Obfervations about
things

things of this kind, I can at prefent find but few among my *Adverfaria*; but in Them I find enough for my prefent turn. For They and my Memory in-form me, that *Vinca per Vinca, Rapha-nus Aquaticus*, Spearemint, and even *Ra-nunculus* it felf, did grow and profper very well in Viols filld with fair water, by whofe Necks the Leaves were fup-ported, and the Plant kept from fink-ing: fome of thefe were onely Cuttings without Roots, divers of them were left in the water all the Autumn, and great part of the Winter, and at the lat-ter end of January were taken out ver-dant, and with fair Roots, which they had fhot in the water. And befides I find, that particularly a Branch or Sprig of *Raphanus Aquaticus* was kept full nine Months, and during that time wither'd not the whole Winter, and was taken out of the water with many fi-brous Roots, and fome green Buds, and an increafe of Weight, and that a Stump

of

of *Ranunculus* did fo profper in the water, that in a Months time it had attain'd
to a pretty deale more then double the
weight it had, when it was put in. And
the next Note, which I find concerning
thefe Plants, informes me, that the a-
bove mention'd Crowsfoot being taken out *agen* at fix Months after it was
put in, weigh'd a Drachm and a half
wanting a Grain and a half, that is, fomewhat above Thrice as much as it did at
firft. This laft Circumftance (of the
increafe of Weight) I therefore thought
fit particularly to make Trial of, and
fet down upon this account among o-
thers, That having doubted the Roots
and Leaves, that feem'd produc'd out of
the Water, might really be fo, by an
Oblongation and an Expanfion of the
Plants, (as I have purpofely try'd, that
an Onion weigh'd and laid up in the
Spring, though after fome weeks keeping in the Air it fhot Blades, whereof
one was five Inches long, in ftead of
incor-

incorporating the Air or terreftrial *Effluviums* with it felf, and confequently thereby growing heavier, had loft nine Grains of its former weight;) it might by this Circumftance appear, that there may be a real Affimilation and Tranfmutation of Water into the Subftance of the Vegetable, as I elfewhere alfo fhew by other proofs. For this being made out, from thence I infer, That the fame Corpufcles, which, convening together after one manner, compofe that fluid, Inodorous, colourlefs, and infipid Body of Water being contexted after other manners, may conftitute differing Concretes, which may have Firmenefs, Opacity, Odours, Smels, Tafts, Colours, and feveral other manifeft Qualities, and that too very different from one another. And befides all this, thefe diftinct Portions of Tranfmuted Water may have many other Qualities, without excepting thofe that are wont to be call'd Specifick, or Occult, witnefs

nefs the feveral Medicinal Virtues attributed by Authors to Spearmint, and to
Periwinckle, to Majorane, and to *Ra-*
phanus Aquaticus. And as for *Ranun-*
culus, that Plant being reckoned among
Poifonous ones, and among thofe that
raife Blifters, 'twill be eafily granted,
that it hath, as other Poifons, an Occult
Deleterial faculty; and indeed it fomewhat deferves our wonder, that fo infipid and innocent a thing as fair Water,
fhould be capable to be turn'd into a
Subftance of fuch a piercing and cauftick
Nature, as by Contact to raife Blifters
on an humane Body. And yet perhaps
that is no leffe ftrange, which we elfewhere relate, That a Plant, confifting
chiefly of Tranfmuted Water, did by
Diftillation afford us a true Oyl, that
would not mingle with Water, and confequently was eafily convertible into
Fire. But whether or no this Experiment, or any fuch like, prove, that almoft All things may be made of All
things,

things, not immediately, but by intervention of fucceffive Changes and Dif-
pofitions, is a Queftion to which we
elfewhere fay fomething, but are not
willing in this place to fay any thing.
And if it be here objected, That the folid Subftance, that accrues to a Plant
rooted in Water, procceds not at all
from the water it felf, but from the Nitrous, fat, and earthy Subftances, that
may be prefum'd to abound even in
common Water, not here to repeat
what I elfewhere fay about this Objection, I fhall at prefent reply, That
though as to divers Plants, that flourifh
after Raine, I am apt to think, as I intimated above, that they may in part be
nourifh'd as well by the Saline and Earthy Subftances, to which the Rain ufually prooves a Vehicle, as by the Rain it
felf; yet as to what the Objection holds
forth about the Plants, that grow not in
the Ground, but in Glaffes fill'd with
Water, it fhould not be barely faid but
prov'd,

prov'd, which he will not perhaps think
eafie to be done, that confiders how
vaft a quantity of fair Water is requifite to be exhal'd away, to obtain as
much as one Ounce of dry Refidents,
whether Saline or Earthy.

III.

That a Plant, growing in the Earth,
doth by the faculties of its Vegetative
Soul attract the Juices of the Earth, that
are within its reach, and felecting thofe
parts that are congruous to its Nature,
refufe the reft, is the general Opinion of
Philofophers, and Phyficians: and therefore many Naturalifts are not wont
much to marvail, when they fee a Tree
bear a Fruit that is fowr or bitter, becaufe they prefume, that Nature hath in
the Root of the Tree cull'd out fuch
parts of the Alimental Juice of the
Earth, as being made to convene into
one Fruit, are fit to make it of fuch a
Quality. But 'tis worth obferving for
our

our prefent purpofe what happens both
in ordinary Graftings, and efpecially in
that kind of Infition (taking the word
in a large fenfe) which is commonly
call'd Inoculation. For though we may
prefume , that the Root of a white
Thorne (for Inftance) may electively
attract its Aliment from the Earth, and
choofe that which is fitteft to produce
the Ignoble fruit, that is proper for that
Plant: yet we cannot reafonably fuppofe, that it fhould in its attraction of
Aliment have any Defigne of providing
an Appropriate Nutriment for a Pere,
and yet the known Experience of Gar-
diners, and our own Obfervations manifeft, that the Cyons of a Pear tree will
take very well upon a White thorn-
ftock, and bring forth a well tafted fruit,
very differing in many qualities from
that of the White thorn.   1 have alfo
learn'd from thofe that are expert, That
though Apples and Pears, being but
Vulgar Fruit, are feldome propagated
                                         but

but by Grafting; yet they may be propagated likewife by Inoculation, (which feems to be but a kind of Grafting with a Bud.) Now in the Inoculations, that are made upon Fruit trees, tis very obfervable, and may much countenance what we are endeavouring to prove, that a little Vegetable Bud, (that is no Seed, properly fo call'd,) not fo big oftentimes as a Pea, fhould be able fo to tranfmute all the Sap that arrives at it, that though this Sap be already in the Root, and in its paffage upwards determin'd by Natures Intention, as Men are wont to fpeak, to the production of the Fruit that is natural to the Stock; yet this Sap fhould by fo fmall a Vegetable Subftance as a Bud, (whether by the help of fome peculiar kind of Strainer, or by the Operation of fome powerful Ferment lodged in it, or by both thefe, or fome other caufe,) be fo far chang'd and overrul'd, as to conftitute a Fruit quite otherwife qualify'd, then that

L which

which is the Genuine production of the
Tree, and which is actually produc'd by
thofe other portions of the like Sap,
which happen'd to nourifh the prolifick
Buds that are the Genuine Of-fpring of
the Stock; fo that the fame Sap, that in
one part of a Branch conftitutes (for inftance) a Clufter of Haws, in another
part of the fame Branch may conftitute
a Pear.   And that which is further remarkable to our prefent purpofe, is,
That not onely the Fruites made of the
fame Sap do often differ from one another in Shape, Bignefs, Colour, Odour,
Taft, and other obvious Qualities, as
well as Occult ones: but that though
the Sap it felf be (oftentimes) a Wate-
rifh and almoft Infipid Liquor, that appears to fenfe Homogeneous enough,
and even by Diftillation affords very
little befides Flegme; yet this Sap is not
onely convertible by Buds of feveral
Natures into differing Fruits, but in one
and the fame Fruit the tranfmuted Sap
                                                        fhall

shall by differing Textures be made to exhibit very differing, and sometimes contrary Qualities. As when (for instance) a Peach bud does not onely change the Sap that comes to it into a Fruit, very differing from that which the Stock naturally produceth, but in the Skin of the Peach it must be red, in the Kernel white, and in other parts of other Colours; the Flesh of it must be fragrant, the Stone inodorous, the Flesh soft and yielding, the Stone very hard and brittle, the Meat pleasantly tasted, the Kernel bitter; not to mention, that Peach Blossoms, though produc'd also by the Bud, are of a Colour and Texture very differing from that of the Fruit, and are enobled with an Occult Quality, which the Fruit hath not, I mean a Purgative Virtue: So that from Inoculations we may learn, That a flegmatick Liquor, that seems Homogeneous enough,& but very slenderly provided with other manifest Qualities then

L 2   common

common water, may, by being varioufly contexted by the Buds of Trees, be tranfmuted into Bodies endow'd with new, and various, and confiderable Sents, Colours, Tafts, Solidity, Medicinal vertues, and divers other Qualities manifeft, and occult.

If it be here faid, that thefe Qualities are the productions of the Plaftick Power refiding in prolifick Buds, which indeed (to me) feem to be but very minute Boughs; I fhall return the fame Anfwer that I did to the like Objection, when 'twas propof'd in the Firft Obfervation.

Hitherto I have onely argued from vulgar Inoculations, but there may be others, as well more confiderable, as leffe ordinary; and i remember I have feen a Tree, whereof, though the Stock was of one fort of good Fruit, there were three more and differing kinds of Stone-fruit, that had been made to take by Inoculation; and two of thofe inoculated

lated Boughs had actually Fruit on
them, and the third, though it had as yet
no Fruit, becaufe the Seafon for that
fort of Plants to bear it was not yet
come, yet the Shoot was fo flourifhing,
that we concluded, that the Bloffoms
would in due time be fucceeded by fruit.
And fince I have been fpeaking of the
differing Qualities of the parts of the
fame Fruit, I am content to adde two
things: the one that *Garcias ab Horto*, a
Claffick Author, (and Phyfician to the
Indian Viceroy) affirmes * with fome
folemnity, (as wondering that a Learned man fhould write otherwife,) that
though the fruit we call *Caffia fiftula* be
very commonly uf'd, both here and in
the *Indies* as a Purging Medicine, yet
the Seeds of this Solutive *Caffia* are
Aftringent. The other: That of late
years there have been often brought
into England from the *Carybbe* Iflands,
certain Kernels of a fruit, which thofe

---

* *Aromat. Hift. lib.* 1. *cap.* 29. *de Caffia folutiva.* *

L 3   that

that have feen it grow, liken to a white Pear-plumme ; thefe are fo ftrongly Purgative, and alfo Emetick, that the Ingenious Mr. *Lygon* * tells us , that five of them wrought with him a Do-zen times upwards, and above Twenty downwards; and yet the fame Author affures us, (which is likewife here a re-ceiv'd Tradition among them that are curious of this fruit,) That in the Ker-nel, in the parting of it into halfes , (as when our Hazle Nuts in *England* part in the middle longwife) you fhall find a thin Filme, which looks of a faint Car-nation, (which colour is eafily enough difcerned, the reft of the Kernel being perfectly white,) and that taking out the Filme you may eat the Nut fafely, without feeling any Operation at all, and 'tis as fweet as a Jordan Almond. [A Learned Man, that practif'd Phyfick in *America*, being inquir'd of by me concerning the Truth of this Relation,

* Ligon's *Hiftory of Barbados*. pag. 67. 68.

an-

anſwer'd, That though he had divers
times given thoſe Nuts as Cathartick
Remedies, yet he had not that Curioſity to take out the Filmes, finding it
the Univerſal belief, that the Purgative
faculty conſiſted therein. And I remember, that the famous * *Monardes*
doth ſomewhat countenance this Tradition, where ſpeaking of another Purging fruit, that alſo comes from *America*,
(from *Cartagena*, and *Nombre de Dios*,)
he takes notice, that theſe purging
Beans (which are like ours, but ſmaller)
have a thin Skin, that divides them
through the middle, which muſt (together with the external Rind) be caſt a-
way, elſe they will work ſo violently
both upwards and downwards, as to
bring the Taker into hazard of his Life:
whereas he commends theſe Beans
rightly prepar'd, not onely as a pleaſant
Medicine, that doth without trouble

---

* See *Nicholaus Monardes*, under the Title, *Fabæ
Purgatrices.*

L 4   puige

purge both Choler, Flegme, and grofs Humors, for which it is celebrated among the *Indians*.

.To thefe ftories of our Countrymen, and *Monardes*, I fhall fubjoin another, which I find related by that great Rambler about the World, *Vincent le Blanck*, who giving us an Account of a publick Garden, which he vifited in *Africa*, in the Territories of the Lord of *Cafama*, not far from the Borders of *Nubia*, which he reprefents as the curiofeft Garden he faw in all the Eaft, he mentions this among other Rarities, " There were (fayes he) other forts of Fruit, which I never faw but there, and one among the reft leav'd like a Sycamore, with fruit like the Golden Apple, but no Gall more bitter, and within five Kernels, as big as Almonds, the Juice whereof is fweet as Sugar, betwixt the Shell and the Nut there grows a thick Skin of a Carnation colour, which

---

Vincent le Blanck's Survey of the World: Part. 2. p. 260.

taken

taken before they be throughly ripe,
they preferve with Date Vinegar, and
make an excellent Sweetmeat, which
they prefent to the King as a great Curiofity.

IV.

The Fourth and laft Obfervation I
fhall at prefent mention, is afforded me
by the confideration of Rotten Cheefe.
For if we take notice of the difference
betwixt two parts of the fame Cheefe,
whereof the one continues found by
preferving its Texture, and the other
hath fuffer'd that Impairing Alteration
of Texture we call Rottennefs, we may
often fee a manifeft and notable Change
in the feveral portions of a Body, that
was before Similar. For the Rotten
part will differ from the Sound in its
Colour, which will be fometimes Livid,
but moft commonly betwixt Green and
Blew; and its *Odour*, which will be both
ftrong and offenfive; and its *Taft*, which
will be very Picquant, and to fome men
much

much more pleafant then before, but to moft men odious; and in divers other Qualities, as particularly its *Confiftence,* it will be much leffe Solid and more Friable then before, and if with a good Microfcope we look upon the moulded parts of many Cheefes, we fhall quickly difcover therein fome Swarms of little Animals, (the Mites,) furnifh'd with *variety* of Parts of differing Sizes, Shapes, Textures, &c. and difcry a yet greater diverfity, both as to manifeft Qualities(nor probably is it inferior as to Occult ones) betwixt the Mouldy part of the Cheefe and the Untainted, then the unaffifted Eye could otherwife have difcovered.

*. *The following Difcourfe* (Of the Origine of Forms) *ought to have been placed before this foregoing Section of the Hiftorical Part.*

OF THE

ORIGINE

OF

FORMS.

❊❊❊❊❊❊❊❊❊❊❊❊❊❊❊❊❊❊❊❊❊❊❊❊❊❊❊❊❊❊❊

THe Origine of Forms, *Pyrophilus*, as it is thought the *Noblest*, so, if I mistake not, it hath been found one of the most *perplex'd* Enquiries, that belong to Natural Philosophy: and, I confesse, it is one of the things that has invited me to look about for some more satisfactory Account, then the Schools usually give of this matter, that I have observ'd, that the wisest that have busied themselves in explicating Forms according to the Peripatetick Notions of them, have either knowingly Confess'd themselves unable to explain them, or unwittingly Prov'd themselves

---

Formarum cognitio est rudis, confusa, nec nisi per ἀναλογίας; neque verum est, forma substantialis speciem recipi in intellectum non enim in sensu usquam fuit. J.C.Scalig.

Formæ substantiales sunt incognitæ nobis, quia insensiles: ideo per qualitates, quæ sunt principia immediata Transmutationis, exprimuntur. Aquinas ad 1. de generat. & corrupt.

In hac humanæ mentis caligine æ quâ forma Ignis ac Magnetis nobis ignota est. Sennertus.

to be fo, by giving but unfatisfactory Explications of them.

It will not (I prefume) be expected, that I, who now write but *Notes*, fhould enumerate, much leffe examine all the various Opinions touching the Origine and Nature of Forms; it being enough for our purpofe, if, having already inti-mated in our *Hypothefis*, what, accor-ding to that, may be thought of this Subject; we now briefly confider the general Opinion of our Modern *Arifto-telians* and the Schools concerning it. I fay, the *Modern* Ariftotelians, becaufe diverfe of the *Antient*, efpecially *Greek* Commentators of *Ariftotle*, feem to have underftood their Mafters Doctrine of Forms much otherwife, and leffe in-congruoufly, then his Latin followers, the Schoolmen and others, have fince done. Nor do I exprefly mention *Ari-ftotle* himfelf among the Champions of fubftantial Forms, becaufe though he feem in a place or two exprefly enough

to

to reckon Formes among *Substances*, yet elsewhere the Examples he imploies to set forth the *Forms* of Natural things by, being taken from the *Figures* of artificial things, (as of a Statue, &c.) which are confessedly but *Accidents*, and making very little use, if any, of Substantial Forms to explain the *Phænomena* of Nature, He seems to me upon the whole matter, either to have been irresolv'd, whether there were any such Substances, or no, or to speak ambiguously and obscurely enough of them, to make it questionable, what his Opinions of them were.

But the summe of the Controversy betwixt Us and the Schools is this, whether or no the Forms of Natural things, (the Souls of Men alwaies excepted) be in Generation *educed*, as they speak, *out of the power of the Matter*, and whether these Forms be true *substantial Entitles*, distinct from the other substantial Principle of Natural Bodies, namely Matter. The

The Reasons that move me to embrace the Negative, are principally these three. *First*, That I see no necessity of admitting in Natural things any such substantial Forms, Matter and the Accidents of Matter being sufficient to explicate as much of the *Phænomena* of Nature, as we either do or are like to understand. *The next*, That I see not what use this puzling Doctrine of substantial Forms is of in Natural Philosophy; the Acute *Scaliger*, and those that have most busied themselves in the Indagation of them, having freely acknowledg'd, (as the more Candid of the Peripateticks generally do,) That the true Knowledg of Forms is too difficult and abstruse to be attain'd by them. And how like it is, that particular *Phænomena* will be explain'd by a Principal, whose Nature is confessedly ignor'd, I leave you to judg: but because to these considerations I often have had, and shall have here and there occasion to

say

fay fomething in the body of thefe Notes, I fhall at prefent infift upon the *third*, which is, That I cannot conceive, neither how Forms can be generated, as the Peripateticks would have it, nor how the things, they afcribe to them, are confiftent with the Principles of true Philofophy, or even with what themfelves otherwife teach.

The Manner how Forms are educed out of the Power of the Matter, according to that part of the Doctrine of Forms, wherein the Schools generally enough agree, is a thing fo Inexplicable, that I wonder not it hath put Acute men upon feveral *Hypothefes* to make it out. And indeed the number of Thefe is of late grown too great to be fit to be here recited, efpecially fince I find them all fo very unfatisfactory, that I cannot but think, the acute Sticklers for any of them are rather driven to embrace it by the palpable inconveniences of the wayes they reject, then by any thing

M they

they find to satisfy them, in that which they make choice of: and for my part I confess, I find so much Reason in what each Party sayes against the Explications of the rest, that I think they all Confute well, and none does well Establish.

But my present way of Writing forbidding me to insist on many Arguments against the Doctrine, wherein they most agree, I shall onely urge, That which I confess chiefly sticks with me, namely that I find it not *Comprehensible.*

I know the Modern Schoolmen fly here to their wonted Refuge of an Obscure Distinction, and tell us, that the Power of Matter in reference to Forms is partly Eductive, as the Agent can make the Form out of it, and partly Receptive, whereby it can receive the Form so made; but since those that say this, will not allow, that the Form of a generated Body was actually præexistent in its Matter, or indeed any where else,

'tis

'tis hard to conceive, how a Subſtance
can be educ'd out of another Subſtance
totally diſtinct in Nature from it, without being, before ſuch Eduction, actually
exiſtent in it. And as for the Receptive Power of the Matter, That but fit-
ting it to receive or lodge a Form, when
brought to be United with it, how can
it be intelligibly made out to contribute
to the Production of a new Subſtance,
of a quite differing Nature from that
Matter, though it harbours it when produc'd? And 'tis plain, that the Humane
Body hath a receptive Power in reference to the Humane Soule, which yet
themſelves confeſs both to be a ſubſtantial Form, and not to be educ'd out
of the Power of Matter. Indeed if they
would admit the Form of a Natural Body to be but a more fine and ſubtle part
of the Matter, as Spirit of Wine is of
Wine, which upon its receſs remains no
longer Wine, but Flegm or Vinegar,
then the Eductive Power of Matter

might

might fignifie fomething ; and fo it
might, if with us they would allow the
Form to be but a Modification of the
Mattter; for then it would import but
that the Matter may be fo order'd or
difpof'd by fit Agents, as to conftitute
a Body of fuch a fort and Denomination: and fo (to refume that Example)
the Form of a Sphære may be faid to
lurk potentially in a piece of Brafs, in as
much as that Brafs may by cafting, turning, or otherwife, be fo figur'd as to
become a Sphære. But *this* they will
not admit, leaft they fhould make Forms
to be but Accidents, though it is for
ought I know as little intelligible, how
what is educ'd out of any Matter, without being either præexiftent, or being
any part of the Matter, can be a true
Subftance, as how that Roundnefs, that
makes a piece of Brafs become a Sphere,
can be a new Subftance in it. Nor can
they admit the *other way* of educing a
Form out of Matter, as Spirit is out of
Wine,

Wine, becaufe then not onely Matter
will be corruptible againft their
grounds, but Matter and Form would
not be two differing and fubftantial
Principles, but one and the fame,
though diverfify'd by firmnefs, and
groffenefs, &c. which are but Accidental differences. I know they fpeak
much of the efficacy of the Agent upon
the Matter, in the Generation of Natural Bodies, and tell us ftrange things
of his manner of working. But not to
fpend time in examining thofe obfcure
niceties, I anfwer in fhort; That fince
the Agent, be he what he will, is but a
Phyfical and finite Agent, and fince
what way foever he works, he can do
noth:ng repugnant to the nature of
things, the difficulty, that fticks with
me, will ftill remain. For if the Form
produc'd in Generation, be, as they
would have it, a Subftance, that was
not before to be found any where out
of that portion of Matter, wherewith it

M 3   con-

conftitutes the Generated Body; I fay
that either it muft be produc'd, by refining or fubtiliating fome parts of the
Matter into Form, or elfe it muft be
produc'd out of nothing, that is, Created, (for I fee no Third way, how a Subftance can be produc'd *de novo.*) If they
allow the Firft, then will the Form be
indeed a Subftance, but not, as they hold
it is, diftinct from Matter; fince Matter,
however fubtiliated, is Matter ftill, as
the fineft Spirit of Wine is as truly a
Body, as was the Wine it felf, that yeil-
ded it, or as is the Groffer Flegm, from
which it was extracted: befides that, the
Peripateticks teach, that the Form is not
made of any thing of the Matter; nor
indeed is it conceivable, how a Phyfical
Agent can turn a Material into an Immaterial Subftance, efpecially Matter
being, as they themfelves confeffe, as
well incorruptible as ingenerable. But
if they will not allow, as indeed they do
not, that the fubftantial Form is made
of

of any thing that is Material, they muſt give me leave to believe, that tis produc'd out of Nothing, till they ſhew me, how a Subſtance can be produc'd otherwiſe, that exiſted no where before. And at this rate every Natural Body of a ſpecial Denomination, as Gold, Marble, Nitre, &c. muſt not be produc'd barely by Generation, but partly by Generation, and partly by Creation. And ſince tis confeſſ'd on all ſides, that no Natural Agent can produce the leaſt Atome of Matter, tis ſtrange they ſhould in Generation allow every Phyſical Agent the power of producing a Form, which, according to them, is not onely a Subſtance, but a far nobler one then Matter, and thereby attribute to the meaneſt Creatures that power of creating Subſtances, which the Antient Naturaliſts thought too great to be aſcrib'd to God himſelf, and which indeed is too great to be aſcrib'd to any other then Him, and therefore ſome Schoolmen

M 4   and

and Philosophers have deriv'd Forms immediately from God; but this is not onely to desert *Aristotle* and the Peripatetick Philosophy they would seem to maintain, but to put Omnipotence, upon working I know not how many thousand Miracles every hour, to performe that (I mean the Generation of Bodies of new Denominations) in a supernatural way, which seems the most familiar effect of Nature in her ordinary course.

And as the Production of Forms out of the Power of Matter is for these Reasons incomprehensible to me, so those things, which the Peripateticks ascribe to their substantial Forms, are some of them such, as, I confesse, I cannot reconcile my Reason to: for they tell us positively, that these Forms are Substances, and yet at the same time they teach, that they depend upon Matter, both *in fieri* and *in esse,* as they speak, so that out of the Matter, that supports them, they

they cannot fo much as exift, (whence they are ufually call'd Material Forms,) which is to make them Subftances in name, and but Accidents in truth: for not to ask how(among Phyfical things) one Subftance can be faid to depend upon another *in fieri*, that is not made of any part of it, the very notion of a Subftance is to be a felf-fubfifting Enti-ty, or that which needs no other Created Being to fupport it, or to make it exift. Befides that, there being but two forts of Subftances, Material, and Immate-rial, a fubftantial Form muft appertain to one of the two, and yet they afcribe things to it, that make it very unfit to be referr'd to either. To all this I adde, that thefe imaginary Material Forms do almoft as much trouble the Doctrine of Corruption, as that of Generation: for if a Form be a true Subftance really diftinct from Matter, it muft, as I lately noted, be able to exift of it felf, with-out any other Subftance to fupport it;

as

as those I reason with confess, that the
Soul of Man survives the Body, it did
before Death inform: whereas they will
have it, that in Corruption the Form is
quite abolish'd, and utterly perishes, as
not being capable of existing, separated
from the Matter, whereunto it was united: so that here again, what they call a
Substance they make indeed an Accident, and besides contradict their own
vulgar Doctrine, That Natural things
are upon their Corruption resolv'd into
the first Matter, since at this rate they
should say, that such things are but partly resolv'd into the first Matter, and
partly either into Nothing, or into
Forms, which being as well immaterial
as the Souls of Men, must, for ought
appears, be also, like them, accounted
immortal.

I should now examine those Arguments, that are wont to be imploy'd by
the Schools to evince their substantial
Forms, but, besides that the nature and
scope

ſcope of my preſent Work injoynes
me Brevity, I confeſſe that, one or two
excepted, the Arguments I have found
mention'd, as the chief, are rather Metaphyſical, or Logical, then grounded upon the Principles and *Phænomena* of
Nature, and reſpect rather Words then
Things, and therefore I, who have neither inclination, nor leaſure, to wrangle
about Terms, ſhall content my ſelf to
propoſe, and very briefly anſwer two or
three of thoſe that are thought the plau-
ſibleſt.

Firſt then they thus argue. *Omne
Compoſitum ſubſtantiale* (for it is hard
to Engliſh well ſuch Uncouth Terms)
*requirit materiam & formam ſubſtantia-
lem, ex quibus componatur. Omne corpus
naturale eſt compoſitũ ſubſtantiale. Ergo*
&c. In this Syllogiſme ſome do plau-
ſibly enough deny the Conſequence,
but for brevities ſake, I ſhall rather
chooſe to deny the Minor, and deſire
the Propoſers to prove it. For I know
not

not any thing in Nature that is compos'd of Matter, and a Substance distinct
from Matter, except Man, who alone is
made up of an immaterial Form, and a
humane Body; and if it be urg'd, that
then other Bodies cannot be properly
said to be *Composita substantialia:* I shall,
rather then wrangle with them, give
them leave to find out some other name
for other Natural things.

But then they argue in the next place,
that, if there were no substantial Forms,
all Bodies would be but *Entia per accidens*, as they speak, which is absurd. To
which I answer, That in the Notion, that
divers Learned men have of an *Ens per
Accidens*, namely, that tis That which
consists of those things, *quæ non ordi-
nantur ad unum*, it may be said, That
though we do not admit substantial
Forms, yet we need not admit Natural
Bodies to be *Entia per accidens*; because
in them the several things that concur
to constitute the Body, as Matter,
Shape,

Shape, Scituation, and Motion, *ordi-*
nantur per fe & intrinfecè to conftitute
one Natural Body. But, if this Anfwer fatisfie not, I fhall adde, that for
my part, That which I am follicitous
about, is, what Nature hath made things
to be in themfelves, not what a Logi-
cian or Metaphyfician will call them in
the Terms of his Art; it being much
fitter in my judgment to alter Words,
that they may better fit the Nature of
Things, then to affix a wrong Nature
to Things, that they may be accommodated to forms of Words, that were
probably devif'd , when the things
themfelves were not known or vvell
underftood, if at all thought on.

Wherefore I fhall but adde one Argument more of this fort, and That is,
that, if there vvere no fubftantial Forms,
neither could there be any fubftantial
Definitions, but the Confequent is abfurd, and therefore fo is the Antecedent.
To vvhich I reply, that fince the Peri-
'pateticks

pateticks themselves confess the Forms
of Bodies to be of themselves unknown, all that this Argument seems to
me to conclude, is but this, *That* if we
do not admit somethings, that are not *in
rerum natura*, we cannot build our Definitions upon them: nor indeed could
we, if we should admit substantial
Forms, give substantial Definitions of
Natural things, unlesse we could also
define Natural Bodies by things that
we know not; for such * the substantial
Forms are (as we have seen already) confess'd to be, by the wisest Peripateticks,
who pretend not to give the substantial
Definition of any Natural *Compositum*,
except Man. But it may suffice Us to
have, instead of *substantial*, *essential* Definitions of things; I mean such as are
taken from the Essential Differences of
things, which constitute them in such a
sort of Natural Bodies, and discriminate

---

* *Nego tibi ullam esse formam nobis notam plenè & planè, nostramq́ scientiam esse umbram in Sole. Scalig.*

them

them from all thofe of any other fort.

Thefe three Arguments, *Pyrophilus*, for fubftantial Forms, You may poffi-bly, as well as I, find varioufly propof'd, and perhaps with fome light alterations multiply'd in the writings of the Peripa-teticks and Schoolmen; but all the Ar-guments of this kind that I have met with, may, if I miftake not, be fuffici-ently folv'd by the Anfwers we have given to thefe, or at leaft by the grounds upon which thofe Anfwers are built; thofe feemingly various Arguments a-greeing in this, That *either* they refpe&t rather Words then Things, *or* that they are grounded upon precarious Suppofi-tions; *or* laftly that they urge That as an Abfurdity, which, whether it be one or not in thofe, that admit the Peripate-tick Philofophy; to me, that do as little acquiefce in many of their other Princi-ples, as I do in their fubftantial Forms, doth not appear any Abfurdity at all. And tis perhaps for fear that Argu-

ments

ments of this fort fhould not much pre-
vaile with Naturalifts, that fome of the
Modern affertors of the Forms we queftion, have thought it requifite to adde
fome more Phyfical Arguments, which
(though I have not found them all in the
fame Writers, yet) being in all but few,
I fhall here briefly confider them.

Firft then among the Phyfical Arguments, that are brought to prove fubftantial Forms, I find That the moft
confidently infifted on, which is taken
from the fpontaneous return of heated
Water to Coldnefs, which Effects, fay
they, muft neceffarily be afcrib'd to the
Action of the fubftantial Form, whofe
office it is to preferve the Body in its
Natural ftate, and, when there is occafion, to reduce it thereunto: and the Argument indeed might be plaufible, if
we were fure, that heated Water would
grow cold again (without the Avolation of any Parts more agitated then the
reft,) fuppofing it to be remov'd into
                                    fome

fome of the imaginary fpaces beyond
the World; but as the cafe is, I fee no
neceffity of flying to a fubftantial
Form, the Matter feeming to be eafily
explicable otherwife. The Water we
heat is furrounded with our Air, or with
fome Veffel, or other Body contiguous
to the Air, and both the Air and the
Water in thefe Climates are moft commonly leffe agitated, then the Juices in
our hands, or other Organs of Touching, which makes us efteem and call
thofe Fluids, cold. Now when the Water is expof'd to the fire, it is thereby
put into a new Agitation, more vehement then that of the parts of our Senfory, which you will eafily grant, if you
confider, that when the Heat is intenfe,
it makes the Water boyl and fmoak,
and oftentimes run over the Veffel; but
when the Liquor is remov'd from the
fire, this acquir'd Agitation muft needs
by degrees be loft, *either* by the avolation of fuch fiery Corpufcles as the

N   Epi-

Epicureans imagine to be got into heated Water, *or* by the Water's communicating the Agitation of its Parts to the contiguous Air, or to the Vessel that contains it, till it have lost its surplusage of Motion, *or* by the ingress of those frigorifick Atoms, wherewith (if any such be to be granted) the Air in these Climates is wont to abound, and so be reduc'd into its former Temperature: which may as well be done without a substantial Form , as if a Ship swimming slowly down a River, should by a sudden gust of Wind, blowing the same way the Stream runs, be driven on much faster then before, the Vessel upon the ceasing of the Wind may, without any such internal principle, return after a while to its former slowness of Motion. So that in this *Phænomenon*, we need not have recourse to an internal principle, the Temperature of the external Air being sufficient to give an account of it. And if Water be kept, (as is usual

fual in poor mens houfes that want Cel-
lars,) in the upper Rooms of the houfe,
in cafe the Climate be hot, the Water
will, in fpight of the Form, continue far
leffe cold, then, according to the Peripateticks, its nature requires, all the
Summer long. And let me here reprefent to the Champions of Forms,
that, according to their Doctrine, the
Fluidity of Water, muft at leaft as
much proceed from its Form as the
Coldneffe, and yet this does fo much
depend upon the Temperature of the
Air, that in *Nova Zembla* vaft quantities of Water are kept in the hard and
folid Form of Ice all the year long, by
the fharp Cold of the ambient Air, notwithftanding all the pretended Office
and Power of the fubftantial Form to
keep it fluid, which it will never be reduc'd to be, unleffe by fuch a thawing
Temperature of the Air, as would it
felf, for ought appears, make it flow a-
gain, although there were no fubftanti-

N 2   all

al Form *in rerum naturà.*

There is another Argument much urg'd of late by some Learned Men, the substance whereof is this; That Matter being indifferent to one sort of Accidents as well as to another, it is necessary there should be a substantial Form, to keep those Accidents, which are said to constitute it, united to the Matter they belong to, and preserve both them and the Body in their Natural state; for since tis confess'd, that Matter hath no appetite to these Accidents, more then to any others, they demand, how without a substantial Form these Accidents can be contain'd and preserv'd? To this I might represent, that I am not so well satisfy'd with the Notion wont to be taken for granted, not onely by the vulgar, but by Philosophers, of the Natural state of Bodies; as if it were undeniable, that every Natural Body, (for as to some, I shall not *now* question it,) has a certain state, wherein Nature endeavours

deavours to preferve it, and out of which it cannot be put, but by being put into a Præternatural ftate. For the World being once conftituted by the great Author of Things, as it now is, I look upon the *Phænomena* of Nature to be cauf'd by the Local Motion of one part of Matter hitting againft another, and am not fo fully convinc'd, that there is fuch a thing, as Natures defigning to keep fuch a parcel of Matter in fuch a ftate, that is cloth'd with juft fuch Accidents, rather then with any other. But I look upon many Bodies, efpecially fluid ones, as frequently changing their ftate, according as they happen to be more or leffe agitated, or otherwife wrought upon by the Sun, and other confiderable Agents in Nature. As the Air, Water, and other Fluids, if the temperature as to Cold or Heat, and Rarefaction or Condenfation, which they are in at the beginning of the Spring here at *London*, be pitcht upon as their Natural ftate,

then

then not onely in the torrid and frozen
Zones they muſt have other and very
differing Natural ſtates, but here it ſelf
they will, almoſt all the Summer and all
the Winter, as our Weather Glaſſes inform us, be in a varying Præternatural
ſtate, becauſe they will be in thoſe ſea-
ſons either more hot and rarify'd, or
more cold and condenſ'd, then in the beginning of the Spring. And in more ſtable and conſtant Bodies I take, in many
caſes, the Natural ſtate to be but either
the moſt uſual ſtate, or *That*, *wherein
that, which produces a notable Change in
them, finds them.* As when a ſlender
piece of Silver, that is moſt commonly
flexible, and will ſtand bent every way,
comes to be well hammer'd, I count
that Flexibility to be the Natural ſtate
of that Mettal, becauſe moſt commonly Silver is found to be flexible, and becauſe it was ſo before it was hammer'd;
but the Springineſſe it acquires by ham-
mering is a ſtate, which is properly no
more

more unnatural to the Silver then the
other, and would continue with the
Mettal as long as It, if both pieces of
Silver, the one flexible, the other fprin-
gy, were let alone, and kept from outward violence: And as the Silver, to be
depriv'd of its flexiblenefs, needed the
violent Motion of the Hammer, fo to
deprive it of its Spring, it needs the violent Agitation of a nealing fire. Thefe
things, and much more, I might here
reprefent, but to come clofe to the Objection, I Anfwer, That the Accidents
fpoken of are introduced into the Matter by the Agents or Efficient Caufes,
whatever they be, that produce in it
what, in the fenfe formerly explain'd, we
call an *effential* (though not a *fubftantial*)
Form. And thefe Accidents being
once thus introduc'd into the Matter,
we need not feek for a new fubftantial
Principle to preferve them there, fince
by the general law, or common courfe
of Nature, the Matter qualify'd by
them,

them, muſt continue in the ſtate ſuch
Accidents have put it into, till, by ſome
Agent or other, it be forcibly put out of
it, and ſo diveſted of thoſe Accidents;
as in the formerly mention'd Example,
borrow'd from *Ariſtotle*, of a Brazen
Sphære , when once the Motion of
Tools, impell'd and guided by the Artificer, have turn'd a piece of Braſs into
a Sphære, there needs no new Subſtance
to preſerve that round figure, ſince the
Braſſe muſt retain it, till it be deſtroy'd
by the Artificer himſelf, or ſome other
Agent able to overcome the reſiſtance
of the Matter, to be put into another
figure. And on this occaſion let me
confirme this *ad hominem*, by repreſen-
ting, That there is not an inconſiderable
Party among the Peripateticks themſelves, who maintain, That in the Elements the Firſt Qualities (as they call
them) are inſtead of Forms , and that
the Fire (for inſtance) hath no other
Form then Heat and Dryneſſe, and the
Wa-

Water then Coldneſſe and Moiſture.
Now if theſe Bodies, that are the vaſt-
eſt and the moſt important of the Sub-
lunary World, confiſt but of the Univerſal Matter, and the few Accidents;
and if in theſe there needs no ſubſtantial Form to keep the Qualities of the
Matter united to it, and conjoyn'd a-
mong themſelves, and preſerve them
in that ſtate, as long as the Law of Nature requires, though beſides the four
Qualities that are call'd *Firſt*, the Elements have divers others, as Gravity
and Levity, Firmneſſe and Fluidity,
Opacouſneſſe and Tranſparency, &c.
why ſhould the favourers of this Opinion deny, That, in other Bodies beſides
the Elements, Qualities may be pre-
ſerv'd and kept united to the Matter
they belong to, without the Band or
Support of a ſubſtantial Form? And as,
when there is no competent deſtructive
Cauſe, the Accidents of a Body will by
the Law of Nature remain ſuch as they
were;

were, fo if there be, it cannot with reafon be pretended, that the fubftantial
Form is able to preferve all thofe Accidents of a Body, that are faid to flow
from it, and to be as it were under its
care and tuition; for if, for inftance, you
expofe a Sphære or Bullet of Lead to a
ftrong fire, it will quickly loofe (not to
mention its Figure) both its Coldnefs,
its Confiftence, its Malleablenefs, its
Colour, (for 'twill appear of the colour
of fire,) its Flexibility, and fome other
Qualities, and all this in fpight of the
imaginary fubftantial Form, which, according to the Peripatetical Principles,
in this cafe muft ftill remain in it without being able to help it. And though
upon the taking the Lead from off the
fire, it is wont to be reduc'd to moft of
its former Qualities, (for it will not of it
felf recover its Sphæricity,) yet That
may well be afcrib'd partly to its peculiar
Texture, and partly to the Coldnefs of
the ambient Air, according to what we
lately

lately difcourf'd tonching heated and re-
frigerated Water, which Temperature
of the Air is an extrinfecal thing to the
Lead, and indeed it is but Accidental,
that the Lead upon refrigeration regains
its former Qualities; for in cafe the Lead
have been expof'd long enough to a fufficiently intenfe fire, it will (as we have
purpofely try'd) be turn'd into Glaffe,
and loofe its colour, its opacity, its malleablenefs, and (former degree of) flexi-
bleneffe, and acquire a Reddifhnefs, a
degree of Tranfparency, a Brittleneffe,
and fome other Qualities, that it had not
before: and let the fuppofed fubftantial
Form do what it can, even when the
Veffel is remov'd from the fire, to reduce or reftore the Body to its Natural
ftate and Accidents, yet the former
Qualities will remain loft, as long as
thefe Præternatural ones, introduc'd by
the fire, continue in the Matter; and neither the one will be reftor'd, nor the o-
ther deftroy'd, till fome fufficiently
power-

powerful extrinfick Agent effect the
Change. And on the other fide I confider, that the Fruit, when fever'd from
the Tree it grew on, is confeff'd to be
no longer animated (at leaft the Kernels,
or Seeds excepted) by the Vegetative
Soul, or fubftantial Form of the Plant;
yet in an Orange or Lemmon (for inftance) pluckt from the Tree, we fee,
that the fame Colour, the fame Odour,
the fame Taft, the fame Figure, the fame
Confiftence, and, for ought we know,
the fame other Qualities, whether fenfible, or even occult, as are its Antido-
tal and Antifcorbutical virtues, that
muft before be faid to have flow'd from
the Soul of the Tree, will continue, many months, perhaps fome years, after
the fruit has ceaf'd to have any com-
merce with the Tree, (nay though the
Tree, whereon it grew, be perhaps in
the mean time hewn down or burnt, and
though confequently its Vegetative
Soul or Form be deftroy'd,) as when it
grew

grew thereon, and made up one Plant with it. And we find, that Tamarinds, Rhubarb, Senna, and many other Simples will for divers years, after they have been depriv'd of their former Vegetative Soul, retain their Purgative and other Specifick properties.

I find it likewise urg'd, that there can be no Reason, why Whiteness should be separable from a Wall, and not from Snow or Milk; unlesse we have recourse to substantial Forms. But in case men have agreed to call a thing by such a name, because it has such a particular Quality, that differences it from others, we need go no farther to find a Reason, why one Quality is essential to one thing, and not to another. As in our former example of a Brass Sphære, the Figure is that, for which we give it that Name, and therefore, though you may alter the figure of the Matter, yet by that very alteration the Body perishes in the capacity of a Sphære, whereas its

Cold-

Coldnefs may be exchang'd for Heat,
without the making it the lefs a
Sphære, becaufe tis not for any fuch
Quality, but for Roundnefs, that a Body is faid to be a Sphære. And fo
Firmnefs is an infeparable Quality
of Ice, though this or that particular
Figure be not, becaufe that tis for want
of fluidity, that any thing, that was immediately before a Liquor, is call'd Ice;
and congruoufly hereunto, though
Whitenefs were infeparable from Snow
and Milk, yet that would not neceffarily
infer, that there muft be a fubftantial
Form to make it fo: for the Firmnefs of
the Corpufcles, that compofe Snow, is as
infeparable from it, as the Whiteneffe;
and yet is not pretended to be the effect
of the fubftantial Form of the Water,
but of the exceffe of the Coldneffe of
the Air, which (to ufe vulgar, though
perhaps unaccurate, expreffions,) puts
the Water out of its Natural ftate of
Fluidity, and into a Præternatural one
of

Firmnefs and Brittlenefs. And the reafon, why Snow feldome loofes its whitenefs but with its nature, feems to be, that its component Particles are fo difpof'd, that the fame heat of the ambient Air, that is fit to turn it into a tranfparent Body, is alfo fit to make it a fluid one, which when it is become, we no longer call it Snow, but Water; fo that the Water loofes its whitenefs, though the Snow do not. But if there be a caufe proper to make a convenient alteration of Texture in the Snow, without melting or refolving it into water, it may then exchange its Whitenefs for Yellowneffe, without loofing its right to be call'd Snow; as, I remember, I have read in an eminent Writer, that *de facto* in the Northern Regions towards the Pole, thofe parcels of Snow, that have lain very long on the ground, degenerate in time into a Yellowifh colour, very differing from that pure Whitenefs to be obferv'd in the neighbouring
Snow

Snow lately fallen.

But there yet remains an Argument for substantial Forms, which though (perhaps because Physical) wont to be overlook'd, or slightly answer'd by their Opposers, will for the same reason deserve to be taken notice of here; and it is, That there seems to be a necessity of admitting substantial Forms in Bodies, that from thence we may derive all the various changes, to which they are subject, and the differing Effects they produce, [the Preservation and Restitution of the State requisite to each particular Body,] as also the keeping of its several parts united into one *Totum*. To the answering of this Argument, so many things will be found applicable, both in the past and subsequent parts of these Notes, that I shall at present but point the chief particulars, on which the Solution is grounded.

I consider then first, that many and great Alterations may happen to Bodies,

dies, which feem manifeſtly to proceed from their peculiar Texture, and the Action of outward Agents upon them, and of which it cannot be ſhewn, that they would happen otherwiſe, though there were no ſubſtantial Forms *in rerum natura*: as we ſee that Tallow (for inſtance) being melted by the fire looſes its Coldneſs, Firmneſs, and its Whiteneſs, and acquires Heat, Fluidity, and ſome Tranſparency, all which, being ſuffer'd to cool, it preſently exchanges for the three firſt nam'd Qualities. And yet divers of theſe Changes are plainly enough the effects partly of the Fire, partly of the ambient Air, and not of I know not what ſubſtantial Form: and it is both evident and remarkable, what great variety of changes in Qualities, and Productions of new ones, the Fire (that is, a Body conſiſting of inſenſible parts, that are variouſly and vehemently mov'd) doth effect by its Heat, that is, *by a modify'd Local Motion.* I conſider

O   further,

further, that various Operations of a
Body may be deriv'd from the peculiar
Texture of the Whole, and the Mechanical Affections of the particular Corpuscles or other parts that compose it,
as we have often occasion to declare here
and there in this Treatise; and particularly by an Instance, ere long to be further insisted on, namely, that though
Vitriol, made of Iron with a Corrosive
liquor, be but a factitious Body, made
by a convenient apposition of the small
parts of the saline Menstruum to those
of the Mettal, yet this Vitriol will do
most, if not all, of the same things, that
Vitriol, made by Nature in the bowels
of the Earth, and digg'd out thence, will
perform; and each of these Bodies may
be endow'd with variety of differing
Qualities, which I see not, why they
must flow, in the native Vitriol, from a
substantial Form, since in the factitious
Vitriol, the same Qualities belong to a
Form, that does plainly emerge from
the

the coalition of Metalline and Saline Corpuscles, associated together and dispos'd of after a certain manner.

And lastly, as to what is very confidently, as well as plausibly, pretended, That a substantial Form is requisite to keep the parts of a Body united, without which it would not be one Body. I answer, That the contrivance of conveniently figur'd parts, and in some cases their juxta-position, may without the assistance of a substantial Form be sufficient for this matter; for not to repeat what I just now mention'd concerning Vitriol made by Art, whose Parts are as well united and kept together, as those of the Native Vitriol, I observe, that a Pear grafted upon a Thorn, or a Plum inoculated upon an Apricock, will bear good fruit, and grow up with the Stock, as though they both made but one Tree, and were animated but by the same common Form; whereas indeed both the Stock and the inoculated or

O 2   graf-

grafted Plant have each of them its own
Form, as may appear by the differing
leaves, and fruits, and feeds they bear.
And that which makes to our prefent
purpofe is, that even Vegetation and
the Diftribution of Aliments are in fuch
cafes well made, though the nourifh'd
parts of the Total Plant, if I may fo call
it, have not one common Soul or Form,
which is yet more remarkable in the
Mifletoes, that I have feen growing upon old Hazletrees, Crab-trees, Apple-
trees, and other plants, in which the
Mifletoe often differs very widely from
that kind of Plant on which it grows
and profpers. And for the durableneffe
of the Union betwjxt Bodies, that a fubftantial Form is not requifite to procure
it, I have been induc'd to think by confidering, that Silver and Gold, being
barely mingl'd by Infufion, will have
their minute parts more clofely united,
then thofe of any Plant or Animal that
we know of. And there is fcarce any
N a-

Natural Body, wherein the Form makes
fo ftrict, durable, and indiffoluble an U-
nion of the parts it confifts of, as that,
which, in that Factitious Concrete we
call Glafs , arifes from the bare com-
miftion of the Corpufcles of Sand with
thofe Saline ones, wherewith they are
colliquated by the violence of the fire:
and the like may be faid of the Union
of the proper Accidents of Glaffe with
the Matter of it, and betwixt one another.

To draw towards a Conclufion, I know
tis alledg'd as a main Confideration on
the behalf of fubftantial Forms, that
thefe being in Natural Bodies the true
principles of their Properties, and confequently of their Operations , their
Natural Philofophy muft needs be very imperfect and defective, who will not
take in fuch Forms: but for my part I
confefs, that this very confideration
does rather indifpofe then incline me to
admit them. For if indeed there were

O 3   in

in every Natural Body fuch a thing as a
fubftantial Form , from which all its
Properties and Qualities immediately
flow, fince we fee that the Actions of
Bodies upon one another are for the
moft part (if not all) immediately perform'd by their Qualities or Accidents,
it would fcarce be poffible to explicate
very many of the explicable *Phænomena*
of Nature, without having recourfe to
Them; and it would be ftrange, if many
of the abftrufer *Phænomena* were not explicable by them onely. Whereas indeed
almoft all the rational Accounts to be
met with of difficult *Phænomena*, are
given by fuch as either do not *acknow-*
ledge, or at leaft do not *take notice* of
fubftantial Forms. And tis evident by the
clear Solutions (untouch'd by many
· vulgar Philofophers,) we meet with of
many *Phænomena* in the Staticks, and o-
ther parts of the Mechanicks, and efpecially in the Hydroftaticks, and Pneu-
maticks, how clearly many *Phænomona*
may

may be folv'd, without imploying a fubftantial Form. And on the other fide,
I do not remember, that either *Ariftotle* himfelf, (who perhaps fcarce ever attempted it,) or any of his Followers,
has given a folid and intelligible folution of any one *Phænomenon* of Nature by
the help of fubftantial Forms; which
you need not think it ftrange I fhould
fay, fince the greateft Patrons of Forms
acknowledg their Nature to be * unknown to Us, to explain any Effect by
a fubftantial Form, muft be to declare
(as they fpeak) *ignotum per ignotius*, or
at leaft *per æquè ignotum*. And indeed
to explicate a *Phænomenon*, being to
deduce it from fomething elfe in Nature more known to Us, then the thing
to be explain'd by It, how can the imploying of Incomprehenfible (or at leaft
Uncomprehended) fubftantial Forms
help Us to explain intelligibly This or

* *Nomina tu lapidis, qui quotidic tuis oculis obfervatur,*
formam, & Phyllida folus habeto. Scal. contra Card.

That

That particular *Phænomenon?* For to say, that such an Effect proceeds not from this or that Quality of the Agent, but from its subftantial Form, is to take an easie way to resolve all difficulties in general, without rightly resolving any one in particular; and would make a rare Philosophy, if it were not far more easie then satisfactory: for if it be demanded, why Jet attracts Straws, Rhubarb purges Choller, Snow dazles the Eyes rather then Graffe, &c. to say, that thefe and the like Effects are perform'd by the fubftantial Forms of the refpective Bodies, is at beft but to tell me, what is the Agent , not how the Effect is wrought; and feems to be but fuch a kind of general way of anfwering, as leaves the curious Enquirer as much to feek for the *caufes* and *manner* of particular Things, as Men commonly are for the particular caufes of the feveral ftrang Things perform'd by Witchcraft , though they be told, that 'tis fome Di-
vel

vel that does them all. Wherefore I do
not think, but that Natural Philofophy,
without being for That the more Defective, may well enough fpare the Doctrine of Subftantial Forms as an ufelefs
Theory; not that Men are yet arriv'd
to be able to explicate all the *Phænome-*
na of Nature without them, but becaufe, whatever we cannot explicate
without them, we cannot neither intelligibly explicate *by* them.

And thus, *Pyrophilus*, I have offer'd
You fome of thofe many things, that in-
difpof'd me to acquiefce in the receiv'd
Doctrine of Subftantial Forms; but in
cafe any more piercing Enquirer fhall
perfwade himfelf, that he underftands it
throughly, and can explicate it clearly,
I fhall congratulate him for fuch happy
Intellectuals, and be very ready to be
inform'd by him. But fince what the
Schools are wont to teach of the Origine and Attributes of fubftantial
Forms, is that, which, I confefs, I cannot

not yet comprehend; and since I have
some of the eminentest Persons among
the Modern Philosophers to joine with
me, though perhaps not for the same
Considerations, in the like confession,
that tis not necessary the Reason of my
not finding this Doctrine conceivable,
must be rather a Defectiveness in my
Understanding, then the unconceivable
nature of the thing it self: I, who love
not (in matters purely Philosophical)to
acquiesce in what I do not understand,
nor to go about to explicate things to
others, by what appears to me it self inexplicable, shall, I hope, be excus'd, if,
leaving those that contend for them, the
liberty of making what use they can of
substantial Forms, I do, till I be better
satisfied, decline imploying them my
self, and endeavour to solve those *Phænomena*, I attempt to give an account of,
without them, as not scrupling to confess, that those that I cannot explicate,
at least in a general way, by intelligible
principles,

principles, I am not yet arriv'd to the diftinct and particular knowledg of.

Now for our Doctrine touching the Origine of Forms, it will not be difficult to collect it from what we formerly difcourf'd about Qualities and Forms together: for the Form of a Natural Body, being according to us, but an Effential Modification, and, as it were, the *Stamp* of its Matter, or fuch a convention of the Bignefs, Shape, Motion (or Reft,) Scituation and Contexture, (together with the thence refulting Qualities) of the fmall parts that compofe the Body, as is neceffary to conftitute and denominate fuch a particular Body; and all thefe Accidents being producible in Matter by Local Motion, tis agreeable to our *Hypothefis* to fay, That the firft and Univerfal, though not immediate caufe of Forms is none other but God, who *put Matter into Motion,* (which belongs not to its Effence,) and *Eftablifh'd the Laws of Motion* amongft Bodies,

Bodies, and also, according to my Opinion, *guided it in divers cases at the beginning of Things*; and that, among Second Causes, the Grand Efficient of Forms is *Local Motion*, which by variously dividing, sequestring, transposing, and so connecting the parts of Matter, produces in them those Accidents and Qualities, upon whose account the portion of Matter they diversifie comes to belong to this or that determinate *species* of Natural Bodies, which yet is not so to be understood, as if *Motion* were onely an Efficient cause in the Generation of Bodies, but very often (as in water, fire, &c.) tis also one of the chiefe *Accidents*, that concurre to make up the Form.

But in this last Summary Account of the Origine of Forms, I think my self oblig'd to declare to you a little more distinctly, what I just now intimated to be my own Opinion. And this I shall do, by advertising you, that though I agree

agree with our *Epicureans*, in thinking
it probable, that the World is made
up of an innumerable multitude of
singly insensible Corpuscles, endow'd
with their own Sizes, Shapes, and Motions; and though I agree with the *Car-*
tesians, in believing (as I find that * *A-*
naxagoras did of Old,) that Matter hath
not its Motion from its self, but Originally from God; yet in This I differ
both from *Epicurus* and *Des Cartes*,
that, whereas the former of them plainly denies, that the World was made by
any Deity, (for Deities he own'd,) and
the Latter of them, for ought I can find
in his Writings, or those of some of his
Eminenteft Disciples, thought, that
God, having once put Matter into Motion, and establish'd the Laws of that
Motion, needed not more particularly

---

* *Aristotle* speaking of *Anaxagoras* in the first Ch. of
the last Book of his *Physicks*, hath this passage: *Dicit*
(*Anaxagoras*) *cùm omnia simul essent, atque quiescerent*
tempore infinito, Mentem movisse, ac segregasse.

interpose

interpose for the Production of Things Corporeal, nor even of Plants or Animals, which according to him are but Engines: I do not at all believe, that either these *Cartesian Laws of Motion,* or the *Epicurean casual Concourse* of Atoms, could bring meer Matter into so orderly and well contriv'd a Fabrick as This World; and therefore I think, that the wise Author of Nature did not onely *put Matter into Motion,* but when he resolv'd to make the World, did so regulate and *guide the Motions* of the small parts of the Universal Matter, as to reduce the greater Systems of them into the Order they were to continue in; and did more particularly contrive some portions of that Matter into Seminal Rudiments or Principles, lodg'd in convenient Receptacles, (and as it were Wombs,) and others into the Bodies of Plants and Animals: one main part of whose Contrivance, did, as I apprehend, consist in this, That some of their Or-

gans

gans were so fram'd, that, suppofing the
Fabrick of the greater Bodies of the U-
niverfe, and the Laws he had eftablifh'd
in Nature, fome Juicy and Spirituous
parts of thefe living Creatures muft be
fit to be turn'd into Prolifick Seeds,
whereby they may have a power, by
generating their like, to propagate their
Species. So that according to my ap-
prehenfion, it was *at the beginning* neceffary, that an Intelligent and Wife
Agent fhould contrive the Univerfal
Matter into the World, (and efpecially
fome Portions of it into Seminal Organs and Principles,) and fettle the
Laws, according to which the Motions
and Actions of its parts upon one another fhould be regulated: without which
interpofition of the Worlds Architect,
however *moving Matter* may with fome
probability (for I fee not in the Notion
any Certainty) be conceiv'd to be able,
after numberlefs Occurfions of its infenfible parts, to caft it felf into fuch.
grand

grand Conventions and Convolutions, as the Cartesians call *Vortices*, and as, I remember, *Epicurus* speaks of under the name of ασομχίσεις, ἠ διηήσεις; yet I think it utterly improbable, that *brute* and *unguided*, though *moving*, Matter, should ever convene into such admirable Structures, as the Bodies of perfect Animals. But the World being once fram'd, and the course of Nature establish'd, the Naturalist, (except in some few cases, where God, or Incorporeal Agents interpose,) has recourse to the first Cause but for its general and ordinary Support and Influence, whereby it preserves Matter and Motion from Annihilation or Desition, and in explicating *particular Phænomena*, considers onely the *Size*, *Shape*, *Motion*, (or *want of it*) *Texture*, and the resulting Qualities and Attributes of the small particles of Matter. And thus in this great *Automaton* the World, (as in a Watch

---

* *Epicurus* in his Epistle to *Pythocles*.

or Clock,) the Materials it confifts of,
being left to themfelves, could never
at the firft convene into fo curious an
Engine: and yet, when the skilful Ar-
tift has once made and fet it a going,
the *Phænomena* it exhibits are to be accounted for by the *number, bigneffe, pro-*
portion, fhape, motion, (or *endeavour,)*
reft, coaptation, and other Mechanical
Affections of the Spring, Wheels, Pil-
lars, and other parts it is made up of:
and thofe effects of fuch a Watch, that
cannot this way be explicated, muft, for
ought I yet know, be confeff'd, not to be
fufficiently underftood.

But to return thither, whence my
Duty to the Author of Nature oblig'd
me, to make this fhort Digreffion.

The hitherto propof'd *Hypothefis,*
touching the Origination of Forms,
hath, I hope, been rendred probable by
divers particulars in the paft Difcourfes, and *will* be both exemplify'd and
confirm'd by fome of the Experiments,

that

that make the Latter part of this prefent Treatife, (efpecially the Fifth and
7th of them,) which, containing Experiments of the Changing the Form of a
Salt and a Mettal, do chiefly belong to
the Hiftorical or Experimental part of
what we deliver touching the Origine
of Forms. And indeed, befides the
two kinds of Experiments prefently to
be mention'd, we might here prefent
you a Third fort, confifting *partly* of
divers Relations of Metalline Tranf-
mutations, deliver'd upon their own
Credit by Credible men, that are not
Alchymifts; and *partly* of fome Experiments (fome made, fome directed by
us) of Changing both Bodies, totally
inflammable, almoft totally into *Water*,
and a good part ev'n of *diftill'd Rain
water* without Additament into *Earth*;
and diftill'd Liquors, readily and totally
mingleable with Water, *pro parte* into
a true *Oyle*, that will not mix with it.
This fort of Experiments, I fay, I might
here

here annex, if I thought fit, in this place,
either to lay any ftreffe upon thofe, that
I cannot my felf make out, or to tranf-
fer hither thofe Experiments of Changes amongft Bodies not Metalline, that
belong to another *Treatife. But o-
ver and above, what the paft Notes and
the Experiments, that are to follow
them, contain towards the making of
what we teach concerning Forms, we
will here, for further Confirmation, proceed to adde two forts of Experiments,
(befides the Third already mention'd.)
The one, wherein it appears, that Bodies
of very differing Natures, being put together, like the Wheels, and other peices of a Watch, and by their connecti-
on acquiring a new Texture, and fo
new Qualities, may, without having recourfe to a fubftantial Form, compofe
fuch a new Concrete, as may as well deferve to have a fubftantial Form attributed to it, by virtue of that new Dif-

* The Sceptical Chymift.

pofition

fition of its parts, as other Bodies that are faid to be endow'd therewith. And the *other*, that a Natural Body being diffipated, and as it were taken in peices, like a Watch, may have its parts fo affociated, as to conftitute New Bodies, of Natures very differing from its own, and from each other; and yet thefe diffipated and fcatter'd parts, by being recollected and put together again, like the pieces of a Watch, in the like order as before, may recompofe (almoft, if not more then almoft) fuch another Body, as that they made up, before they were taken afunder.

EXPE-

I.

EXPERIMENTS,

and THOUGHTS, about the *Production* and *Reproduction* of FORMS.

IT was not at randome, that I spoke, when, in the foregoing Notes about the Origine of Qualities, I intimated, That 'twas very much by a kind of tacit agreement, that Men had diftinguifh'd the *Species* of Bodies, and that thofe Diftinctions were more Arbitrary then we are wont to be aware of. For I confeffe, that I have not yet, either in *Ariftotle*, or any other Writer, met with any genuine and fufficient Diagnoftick and Boundary, for the Difcriminating

and

and limiting the *Species* of Things, or to
speak more plainly, I have not found,
that any Naturalist has laid down a determinate Number and sort of Qualities, or other Attributes, which is *sufficient* and *necessary* to constitute all portions of Matter, endow'd with them, distinct Kinds of Natural Bodies. And
therefore I observe, that most commonly Men look upon these as Distinct *Species* of Bodies, that have had
the luck to have distinct Names found
out for them; though perhaps diverse
of them differ much lesse from one another, then other Bodies, which (because
they have been hudled up under one
Name,) have been look'd upon, as but
one sort of Bodies. But not to lay any
weight on this Intimation about
Names, I found, that for want of a true
Characteristick, or discriminating notes,
it hath been, and is still, both very *un-
certain* as to divers Bodies, whether they
are of different *Species* or of the same,
and

and very *difficult* to give a fufficient reafon, why divers Bodies, wherein Nature
is affifted by Art, fhould not as well pafs
for diftinct kinds of Bodies, as others,
that are generally reckon'd to be fo.

Whether (for inftance) Water and
Ice be not to be efteem'd diftinct kinds
of Bodies, is fo little evident, that fome,
that pretend to be very well verf'd in
Ariftotle's Writings and Opinions, affirme him to teach, that Water loofes
not its own nature by being turn'd into
Ice; and indeed I remember I have read
a * Text of his, that feems exprefs e-
nough to this purpofe, and the thing it
felf is made plaufible by the reducible-
neffe of Ice back again into Water. And
yet I remember , *Galen* is affirm'd to
make thefe two, diftinct *species* of Bo-

* See *Lib. 1. de Gen. & Cor. t. 80.* Idem Corpus (fayes he
there) *quamquam continuum, alias liquidum, alias concretum videmus, non divifione aut compofitione hoc paffum, aut
converfione, aut attactu, ficuti Democritus afferit: nam
neque transpofitione, neque Naturæ demutatione (ὲ τὸ μὲ-
ταϛάλλον τὴν φύσιν) ex liquido concretum evadere folet.*

dies;

dies; which Doctrine is favour'd by the differing Qualities of Ice and Water, for not onely the one is fluid, and the other solid, and even brittle, but Ice is also commonly more or less opacous in comparison of Water, being also lighter then it *in specie*, since it swims upon it. To which may be added, that Ice, beaten with common Salt, will freez other Bodies, when Water mingled with Salt will not. And on this occasion, I would propose to be resolv'd, whether Must, Wine, Spirit of Wine, Vinegar, Tartar, and Vappa, be Specifically distinct Bodies? and the like question I would ask concerning a Hens Egg, and the Chick that is afterwards hatch'd out of it: As also concerning Wood, Ashes, Soot, and likewise the Eggs of Silkworms, which are first small Caterpillars, or (as some think them) but Worms, when they are newly hatch'd, and then *Aurelia's*, (or husked Maggots,) and then Butterflies, which I

have

have obferv'd with pleafure to be the
fucceffive Production of the Prolifick
Seed of Silkworms. And whether the
Anfwer to thefe Quæries be Affirmative
or Negative, I doubt the reafon, that
will be given for either of the two, will
not hold in divers cafes, whereto I might
apply it. And a more puzling Queftion it may be to fome, whether a Char.
coal, being throughly kindled, do fpecifically differ from another Charcoal?
for, according to thofe I argue with, the
fire has *penetrated* it quite through; and
therefore fome of the recent *Ariftoteli-*
ans are fo convinc'd of its being tranfmuted, that all the fatisfaction i could
find from a very fubtle modern School-
man to the Objection, That if the glowing Coal were plung'd into Water, it
would be a black Coal agen, was, That
notwithftanding That reduction, the
Form of a Charcoal had been once a-
bolifh'd by the fire, and was reproduc'd
by God, upon the regain'd Difpofition
of

of the Matter to receive it.

Nor is it very easie to determine, whether Clouds, and Rain, and Hail, and Snow, be bodies specifically distinct from Water, and from each other, and the writers of Meteors are wont to handle them as distinct. And since if such slight differences as those, that discriminate these Bodies, or that which distinguishes Wind from Exhalations, whose Course makes it, be sufficient to constitute differing kinds of Bodies, 'twill be hard to give a satisfactory Reason, why other Bodies, that differ in more or more considerable particulars, should not enjoy the same Priviledge. And I presume, that Snow differs less from Rain, then Paper doth from Rags, or Glass made of Wood-ashes does from Wood. And indeed, Men having, by tacit consent, agreed to look upon Paper, and Glass, and Soape, and Sugar, and Brass, and Ink, and Pewter, and Gunpowder, and I know not how many others, to be diftinct

ſtinct ſorts of Bodies, I ſee not, why they
may not be thought to have done it, on
as good grounds, as thoſe, upon which
divers other diſtering *Species* of Bodies
have been conſtituted. Nor will it ſuffice to object, that theſe Bodies are *factitious*; for 'tis the preſent nature of
Bodies, that ought to be conſider'd in
referring them to *Species*, which way
foever they came by that Nature: for
Salt, that is, in many Countries, made by
boiling Sea water in Cauldrons, and o-
ther veſſels, is as well true Sea-ſalt, as
that which is made in the Iſle of *Man*,
(as Navigators call it,) without any co-
operation of Man, by the bare action of
the Sun upon thoſe parts of the Sea water, which chance to be left behind in
hollow places, after a high Spring-tide.
And Silk-worms, which will hatch by
the heat of humane Bodies, and Chickens, that are hatch'd in *Ægypt* by the
heat of Ovens or Dunghils, are no leſs
true Silk-worms or Chickens, then
                                    thoſe

thoſe that are hatch'd by the Sun, or by Hens.

As for what may be objected, that we muſt diſtinguiſh betwixt Factitious Bodies and Natural, I will not now ſtay to examine, how far that Diſtinction may be allow'd: for it may ſuffice for our preſent purpoſe to repreſent, that whatever may be ſaid of Factitious Bodies, where Man does, by Inſtruments of his own providing, onely give Figure, or alſo Contexture to the *ſenſible* (not infenfible) parts of the Matter he works upon; as when a Joyner makes a Stool, or a Statuary makes an Image, or a Turner a Bowl: yet the caſe may be very differing in thoſe other factitious Productions, wherein the *infenſible* parts of Matter are alter'd by Natural Agents, who perform the greateſt part of the work among themſelves, though the Artificer be an Aſſiſtant, by putting Them together after a due manner. And therefore I know not, why all the Productions

ductions of the Fire made by Chymifts
fhould be look'd upon, as not Natural,
but Artificial Bodies: fince the Fire,
which is the grand Agent in thefe
Changes, doth not, by being imploy'd
by the Chymift, ceafe to be, and to
work as, a Natural Agent. And fince
Nature her felf doth, by the help of the
fire, fometimes afford us the like Productions that the Alchymifts art pre-
fents us: as in *Ætna, Vefuvius*, and o-
ther burning Mountains, (fome of whofe
Productions I can fhew you,) Stones
are fometimes turn'd into Lime, (and fo
an Alcalizate Salt is produc'd,) and
fometimes, if they be more difpof'd to
be flux'd, then calcin'd, brought to vitrification; Metalline and Mineral Bodies are by the violence of the fire colliquated into Maffes of very ftrange and
compounded Natures. Afhes and Metalline flowers of divers kinds are fcatter'd about the neighbouring places, and
copious flowers of Sulphur, fublim'd
by

by the internal fire, have been several times found about the Vents, at which the Fumes are discharg'd into the Air: (As I have been assur'd by Ingenious Visiters of such Places, whom I purposely inquir'd of, touching these *flores*; for of these Travellers more then one answer'd me, they had themselves gather'd, and had brought some very good.) Not to adde, that I have sometimes suspected, upon no absurd grounds, that divers of the Minerals and other Bodies, we meet with in the lower parts of the Earth, and think to have been formed and lodg'd there ever since the beginning of Things, have been since produc'd there by the help of subterraneal fires, or other heats, which may *either* by their immediate action, and exceedingly long application, very much alter some Bodies by changing their Texture; as when Lead is turn'd into Minium, and Tin into Putty by the operation of the fire in a few hours, *or* by ele-

elevating, in the form of Exhalations or
Vapours, divers Saline and Sulphureous
Corpuscles or Particles of unripe (or to
use a Chymical Term of Art) Embrio-
nated Minerals, and perhaps Mettals,
which may very much alter the Nature,
and thereby vary the Kind of other subterraneal Bodies, which they pervade,
and in which they often come to be in-
corporated; *or else* may, by convening
among themselves, constitute particular Concretions, as wee see that the
fumes of Sulphur and those of Mercury unite into that Lovely red Mass,
which in the Shops they call Vermilion,
and which is so like to the Mineral,
whence we usually obtain Mercury, that
the *Latines* give them both the same
Name *Cinnabaris*, and in that are imitated by the French and italians; in
whose favour I shall adde, That if we sup-
pose this Mineral to consist of a stony
Concretion, penetrated by such Mineral fumes as I have been speaking of, the
Ap-

Appellation may be better excuf'd then
perhaps you imagine, fince from *Cinna-*
baris nativa not onely I obtain'd a confiderable quantity of good running
Mercury, (which is That, Men are wont
to feek for from it,) but to gratifie my
Curiofity fomewhat turther, I try'd an
eafie way, that came into my mind,
whereby the *Caput mortuum* afforded
me no defpicable Quantity of good
combuftible Sulphur. But this upon
the By, being not oblig'd to fet down
here the grounds of my Paradoxical
Conjecture about the Effects of fubter-
raneal Fires and Heats, fince I here lay
no ftrefs upon it, but return to what I
was faying about *Ætna*, and other Vol-
cans. Since then thefe Productions of
the Fire, being of Nature's own make-
ing, cannot be deny'd to be Natural
Bodies, I fee not why the like Productions of the Fire fhould be thought
unworthy that Name, onely becaufe
the Fire, that made the former, was kind-

led

led by chance in a Hill, and that which produc'd the latter was kindled by a Man in a Furnace. And if flower of Sulphur, Lime, Glafs, and colliquated mixtures of Metals and Minerals are to be reckon'd among Natural Bodies, it feems to be but reafonable, that, upon the fame grounds, we fhould admit flower of Antimony, Lime, and Glafs, and Pewter, and Brafs, and many other Chymical Concretes, (if I may fo call them) to be taken into the fame number; and then 'twill be evident, that to diftinguifh the *fpecies* of Natural Bodies, a Concourfe of Accidents will, without confidering any Subftantial Form, be fufficient.

But becaufe I need not, on this occafion, have recourfe to inftances of a difputable nature, I will pitch, for the illuftration of the Mechanical Production of Forms, upon Vitriol. For fince Nature her felf, without the help of Art, does oftentimes produce that Concrete,

Q   (as

(as I have elfewhere fhewn by Experience,) there is no reafon why Vitriol,
produc'd by eafie Chymical Operations, fhould not be look'd upon as a Body of the fame Nature and Kind. And
in Factitious Vitriol, our knowing what
Ingredients we make ufe of, and how
we put them together, inables us to
judge very well, how Vitriol is produc'd. But becaufe it is wont to be
reckon'd with Salt-petre, Sea-falt, and
Sal Gem among true Salts, I think it requifite to take notice in the firft place,
that Vitriol is not a meer Salt, but That,
which *Paracelfus* fomewhere, and after
him divers other Spagyrifts, call a Magiftery, which in their fenfe (for there
are that ufe it in another,) commonly
fignifies a Preparation, wherein the
Body to be prepar'd has not its *Princi-*
ples feparated, as in Diftillation, Incineration, &c. but wherein the *whole Body*
is brought into another form, by the
addition of fome Salt or Menftruum,
that

that is united *per minima* with it. And
agreeably to this Notion we find, that
from common Vitriol, whether native
or factitious, may be obtain'd (by Diftillation and Reduction) an acid Saline
Spirit, and a Metalline Subftance, as I
elfewhere mention, that from blew Vitriol, Copper may be (by more then
one way) feparated. And I the rather
give this Advertifement, becaufe that
as there is a Vitriol of Iron, which is u-
fually green; and another of Copper,
which is wont to be blew; and alfo a
white Vitriol, about which it is difputed
what it holds, (though that it holds
fome Copper I have found,) and yet all
of thefe are without fcruple reputed true
Vitriols, notwithftanding that they differ fo much in Colour, and (as I have
difcover'd) in feveral other Qualities; fo
I fee no reafon, why the other Minerals,
being reduc'd by their proper Menftruums into Salt like Magifteries, may not
pafs for the Vitriols of thofe Metals,

Q 2   and

and confequently for Natural Bodies:
which, if granted, will adde fome confirmation to our Doctrine, though its
being granted is not neceffary to make it
out. For, to confine our felves to Vitriol, 'tis known among Chymifts, that
if upon the filings of *Mars* one put a
convenient quantity of that acid diftill'd
Liquor, which is (abufively) wont to
be call'd Oyl of Vitriol, diluting the
mixture with Rain, or with common
Water, 'tis eafie by Filtrating the Solution, by Evaporating the Aqueous
fuperfluity of it, and by leaving the reft
for a competent while in a Cellar, (or
other cold place) to Chriftallize, 'tis
eafie, I fay, by this means to obtain a
Vitriol of Iron; which agrees with the
other Vitriol of Vitriol-ftones or Mar-
chafites, prefented us, by Nature, without the help of any other Menftruum,
then the Rain that falls upon them from
the Clouds, in I know not how many
Qualities, part Obvious, and part of
them

them Occult: As, (of the *first sort*) in Colour, Tranſparency, Brittleneſſe, eaſineſs of Fuſion, Styptical Taſt, reduciblenels to a Red Powder by Calcination, and other Qualities more obvious to be taken notice of; to which may be annex'd divers Qualities of the *ſecond sort,* (I mean the more abſtruſe ones,) as the power to turn in a trice an Infuſion of Galls, made in ordinary water, (as alſo to turn a certain clear Mineral Solution, elſewhere mention'd,) into an Inckly colour, to which, in all probability, we may adde a faculty of cauſing Vomits even in a ſmall Doſe, when taken into the Stomach of a Man, and that remarkable property of being endow'd with as exact and curious a ſhape or figure, as Thoſe , for which Salts have been , by modern Philoſophers eſpecially, ſo much admir'd. But, that no ſcruple might ariſe from hence, that in the *Vitriolum Martis,* wont to be made by Chymiſts, the Menſtruum,

Q 3   that

that is imploy'd, is the Oyl of common Vitriol, which may be fuspected to have retain'd the nature of the Concrete whence it proceeded, and fo this Factitious Vitriol may not be barely a new Production, but partly a Recorporification, as they fpeak, of the Vitriolate Corpufcles contain'd in the Menftruum: To prevent this Scruple I fay, (which yet perhaps would not much trouble a Confidering Chymift,) I thought fit to imploy a quite other Menftruum, that would not be fufpected to have any thing of Vitriol in it. And though *Aqua fortis*, and Spirit of Nitre, however they *corrode* Mars, are unfit for fuch a work, yet having pitch'd upon Spirit of Salt inftead of Oyl of Vitriol, and proceeding the fame way that has been already fet down, it anfwer'd our Expectation, and afforded us a good green Vitriol. Nor will the great difpofition, I have obferv'd in this our Vitriol, to refolve, by the moifture of the Air, into

a

à Liquor, make it effentially differing from other Vitriols, fince it has been obferv'd, and particularly by *Guntherus Belichius* more then once, that even the common Vitriol he uf'd in *Germany*, will alfo, though not fo eafily as other Salts, run (as the Chymifts phrafe it) *per deliquium*. And to make the Experiment more compleat, though we did not find either Oyl of Vitriol, or Spirit of Salt, good Menftruums to make a blew Venereal Vitriol out of Copper, (however fil'd, or thinly laminated,) and though upon more Tryals then one, it appear'd, that *Aqua fortis*, & Spirit of Nitre, which we thought fit to fubftitute to the above mention'd Liquors, did indeed make a Solution of Copper, but fo unctuous a one, that twas very hard to bring any part of it to dryneffe, without fpoyling the Colour and Shape of the defir'd Body: yet repeating the Experiment with care and watchfulnefs, we, this way, obtain'd one of the lovelieft lovelieſt Vitriols that hath perhaps been ſeen, and of which you your ſelf may be the judg by a parcel of it I keep by me for a Rarity.

To apply now theſe Experiments, eſpecially That, wherein Spirit of Salt is imploy'd, to the purpoſe, for which I have mention'd them, let us briefly conſider theſe two things; the one, that our Factitious Vitriol is a Body, that, as well as the Natural, is endow'd with many Qualities, (manifeſt, and occult,) not onely ſuch as are common to it with other Salts, as Tranſparency, Brittleneſs, Solublenesſe in Water, &c. but ſuch as are Properties peculiar to it, as Greenneſs, eaſineſs of Fuſion, Stypticity of Taſt, a peculiar Shape, a power to ſtrike a Black with infuſion of Galls, an Emetick faculty, &c.

The other thing we are to conſider is, that though theſe Qualities are in common Vitriol believ'd to flow from the ſubſtantial Form of the Concrete, and

and may, as juſtly as the Qualities, whether manifeſt or occult, of other Inanimate Bodies, be imploy'd as Arguments
to evince ſuch a Form: yet in our Vitriol, made with Spirit of Salt, the ſame
Qualities and Properties were produc'd
by the aſſociating and juxtapoſition of
the two Ingredients, of which the Vitriol was compounded, the Myſtery
being no more but this, That the Steel
being diſſolv'd in the Spirit, the Saline
Particles of the former, and the Metalline ones of the latter, having each their
Determinate Shapes, did by their Aſſo-
ciation compoſe divers Corpuſcles of a
mix'd or compounded Nature, from the
Convention of many whereof, there re-
ſulted a new Body, of ſuch a Texture,
as qualify'd it to affect our Senſories,
and work upon other Bodies, after ſuch
a manner as common Vitriol is wont to
do. And indeed in our caſe, not onely it cannot be made appear, that there
is any ſubſtantial Form generated anew,
but

but that there is not so much as an exquisite mixture, according to the common Notion the Schools have of such a Mixture. For Both the Ingredients retain their Nature, (though perhaps somewhat alter'd,) so that there is, as we were saying, but a Juxta-position of the Metalline and Saline Corpuscles; onely they are associated so, as by the mannner of their Coalition to acquire that new Texture, which Denominates the Magistery they compose, Vitriol. For 'tis evident, that the Saline Ingredient may either totally, or for much the greatest part be separated by Distillation, the Metalline remaining behind. Nay some of the Qualities, we have been ascribing to our Vitriol, do so much depend upon Texture, that the very Beams of the Sun (converg'd) will, as I have purposely try'd, very easily alter its Colour, as well as spoyl its Transparency, turning it at first from Green to White, and, if they be concenter'd

center'd by a good Burning glaſs, make-
ing it change that Livery for a deep Red.

Doubts and Experiments, touching the
Curious Figures of SALTS.

ANd here let me take notice, that
though the exact and curious Figures, in which Vitriol and other Salts
are wont to ſhoot, be made Arguments
of the *Preſence*, and great Inſtances of
the *Plaſtick skill* of ſubſtantial Forms
and Seminal Powers, yet, I confeſs, I
am not ſo fully ſatisfied in this matter,
as even the Modern Philoſophers appear to be. Tis not that I deny, that
Plato's excellent Saying, ποιμταῖ ὁ Θεός,
may be apply'd to theſe exquiſite Productions of Nature. For though God
has thought fit to make things Corporeal after a much more facile and intelligible way, then by the intervention of
ſubſtantial Forms; and though the Pla-
                                    ſtick

ſtick power of Seeds, which in Plants
and Animals I willingly admit, ſeem not
in our caſe to be needful; yet is the Divine Architect's Geometry (if I may ſo
call it)neverthelefſe to be acknowledg'd
and admir'd; for having been pleaſ'd to
make the *primary* and infenſible Corpuſcles of Salts and Metals of ſuch determinate, curious, and exact Shapes,
that, as they happen to be affociated together, they ſhould naturally produce
Concretions, which, though *differingly* figur'd according to the refpective
Natures of their Ingredients, and the
various manners of their Convening,
ſhould yet be all of them very *curious*,
and ſeem elaborate in their Kinds. How
little I think it fit to be allow'd, that the
Bodies of Animals, which conſiſt of ſo
many curiouſly fram'd and wonderfully
adapted Organical parts, (and whoſe
Structure is a thouſand times more Artificial then that of Salts, and Stones,
and other Minerals,) can be reaſonably
ſup-

suppos'd to have been produc'd by Chance, or without the Guidance of an Intelligent Author of Things, I have elsewhere largely declar'd. But I confess, I look upon these Figures we admire in Salts, and in some kinds of Stones, (which I have not been Incurious to collect,) as Textures so simple and slight in comparison of the Bodies of Animals, & oftentimes in comparison of some one Organical part, that I think it cannot be in the least inferr'd, that because such slight Figurations need not be ascrib'd to the Plastick power of Seeds, it is not necessary, that the stupendious and incomparably more elaborate Fabrick and structure of Animals themselves should be so. And this premis'd, I shall adde, that I have been inclin'd to the Conjecture about the shapes of Salts, that I lately propos'd, by these Considerations.

First, That by a bare Association of Metalline and Saline Corpuscles, a Concrete

crete, as finely figur'd as other Vitriols, may be produc'd, as we have lately seen.

Secondly, becaufe that the Figures of thefe Salts are not conftantly in all refpects the fame, but may in diverfe manners be fomewhat varied, as they happen to be made to fhoot more haftily, or more leifurely, and as they fhoot in a fcanter, or in a fuller proportion of Liquor. This may be eafily obferv'd by any, that will but with a little Attention confider the difference that may be found in Vitriolate Chriftals or Grains, when quantities of them were taken out of the great Coolers, as they call them, wherein that Salt, at the Works where tis boyl'd, is wont to be fet to fhoot. And accordingly, where the Experienc'd Mineralift *Agricola*, defcribes the feveral wayes of making Vitriol in great Quantities, he does not onely more then once call the great Grains or Chriftals, into which it coagulates, Cubes; but fpeaking of the manner of their

Con-

Concretion about the Cord's or Ropes,
that are wont (in *Germany*) to be hang'd
from certain crofs Bars into the Vitriolate Water or Solution for the Vitriol
to faften its felf to; he compares the
Concretions indifferently to Cubes or
Clufters of Grapes: *Ex his* (fayes he,
fpeaking of the crofs Bars,) *pendent re-*
ftes lapillis extentæ, ad quos humor fpif-
fus adhærefcens denfatur in tranflucentes
atramenti futorii vel Cubos, vel Acinos,
qui uvæ fpeciem gerunt. I remember alfo, that having many years fince a fufpicion, that the Reafon why *Alkalys*,
fuch as Salt of Tartar and Pot-afhes
are wont to be obtain'd in the form of
white Powders or *Calces*, might be the
way, wherein the Water, or the Lixiviums, that contain them, is wont to be
drawn off, I fancied, that by leaving the
Saline Corpufcles a competent quanti-
of Water to fwimme in, and allowing
them leafure for fuch a multitude of

* *Georg: Agricola de re metall. lib. 12.p. 462.*

Occur-

Occurfions , as might fuffice to make
them hit upon more congruous Co-
alitions then is ufual , I might obtain
Chriftals of Them, as well as of other
Salts: conjecturing this, I fay, I cauf'd
fome well purify'd *Alkalys* , diffolv'd in
clear water, to be flowly evapourated,
till the Top was cover'd with a thin Ice-
like Cruft, then taking care not to break
That, leaft they fhould (as in the ordinary way, where the Water is all forc'd
off,) want a fufficient ftock of Liquor,
I kept them in a very gentle heat for a
good while; and then breaking the a-
bove mentioned Ice-like Cake, I had, as
I wifh'd, divers figured Lumps of Chriftalline Salt fhot in the Water, and
tranfparent almoft like white Sugar
Candy.

I likewife remember, that having, on
feveral occafions, diftill'd a certain quantity of Oyl of Vitriol, with a ftrong Solution of Sea-falt , till the remaining
Matter was left dry, that Saline Refidue

due being diſſolv'd in fair water, filter'd, and gently evaporated, would ſhoot in-to Chriſtals, ſometimes of one figure, ſometimes of another, according as the quantity or ſtrength of the Oyl of Vitriol and other Subſtances determin'd. And yet theſe Chriſtals, though ſometimes they would ſhoot into Priſme-like Figures , as Roch'd Petre; and ſometimes into ſhapes more like to Allome or Vitriol; nay though oftentimes the ſame *Caput mortuum* diſſolv'd, would in the ſame Glaſs ſhoot into Chriſtals, whereof ſome would be of one ſhape, ſome of another; yet would theſe differing Grains or Chriſtals appear for the moſt part more exquiſitely figur'd, then oftentimes Vitriol does. From Spirit of Urine and Spirit of Nitre, when I have ſuffer'd them to remain long together before Coagulation, and free'd the mixture from the ſuperfluous moiſture very ſlowly, I have ſometimes obtain'd fine long Chriſtals,

R   (ſome

(some of which I can shew you) so shap'd, that most Beholders would take them for Christals of Salt-petre. And I have likewise tryed, that whereas Silver is wont to shoot into Plates exceeding thin, almost like those of *Moscovia* glass, when I have dissolv'd a pretty quantity of it in *Aqua fortis*, or spirit of Nitre, and suffer'd it to shoot very leisurely, I have obtain'd Lunar Christals, (several of which I have yet by me) whose Figure, though so pretty as to have given some wonder even to an Excellent Geometrician, is differing enough from that of the thin Plates formerly mention'd; each Christal being compos'd of many small and finely shap'd Solids, that stick so congruously to one another, as to have one surface, that appear'd Plain enough, common to them all.

Thirdly, that insensible Corpuscles of different, but all of them exquisite, shapes, and endowed with plain as well

as

as fmooth fides, will conſtitute Bodies
varioufly, but all very finely figur'd, I
have made uſe of feveral waies to manifeſt. And firſt, though Harts-horn,
Bloud, and Urine, being refolv'd, and
(as the Chymiſts ſpeak) Analiz'd by
Diſtillation, may well be ſuppoſ'd to
have their ſubſtantial Forms (if they
had any) deſtroy'd by the action of the
Fire: yet in regard the Saline Particles,
they contain, are endow'd with ſuch figures as we have been ſpeaking of, when
in the Liquor, that abounds with either
of theſe volatile Salts, the diſſolv'd
Particles do leiſurely ſhoot into Chriſtals, I have divers times obſerv'd, in
theſe, many Maſſes, (ſome bigger, and
ſome leſs,) whoſe ſurfaces had Plains,
ſome of Figures, as to ſenſe exactly Geometrical, and others very curious and
pleaſant. And of theſe finely ſhap'd
Chriſtals of various ſizes, I have pretty
ſtore by me. And becauſe (as it may
be probably gather'd from the Event)

R 2   the

the Saline Corpuscles of Stillatitious acid liquors, and those of many of the Bodies, they are fitted to diffolve, have fuch kind of Figures as we have been fpeaking of, when the folutions of thefe Bodies, upon the recefs of the fuperfluous moifture, fhoot into Chriftals; thefe, though they will fometimes be differing enough, according to the particular natures of the diffolv'd Bodies and the Menftruum, yet either the Chriftals themfelves, or their Surfaces, or both, will oftentimes have fine and exquifite Figures; as I have try'd by a Menftruum, wherewith I was able to diffolve fome Gems; as alfo with a folution of Coral, made with Spirit of Verdigreefe, to omit other Examples. And for the fame reafon, when I try'd whether the Particles of Silver, diffolv'd in *Aqua fortis*, would not, without Concoagulating with the Salts, convene, upon the Account of their own fhapes, into little Concretions of fmooth and flat furfaces,

ces, I found, that having (to afford the
Metalline Corpuscles scope to move
in) diluted one part of the Solution
with a great many parts of distill'd Rain
water, (for common water will oftentimes make such Solutions become
white or turbid,) a Plate of Copper being suspended in the Liquor, and suffer'd
to lie quiet there a while, (for it need
not be long)there would settle, all about
it, swarms of little Metalline and Undia-
phanous Bodies, shining in the water
like the scales of small Fishes, but form'd
into little Plates extremely thin, with
surfaces not onely flat, but exceeding
glossy: and among those, divers of the
larger were prettily figur'd at the
Edges. And as for Gold, its Corpuscles are sufficiently dispos'd to convene
with those of fit or congruous Salts into
Concretions of determinate Shapes, as
I have found in the Christals I obtain'd
from Gold dissolv'd in *Aqua Regis*, and
after having been suffer'd to loose its

super-

superfluous moisture, kept in a cold place: and not onely so, but also when by a more powerful Menstruū I had subdivided the Body of Gold into such minute Particles, that they were sublimable, (for That, I can assure you, is possible,) these volatile Particles of Gold, with the Salts, wherewith they were elevated, afforded me (sometimes) store of Christals, which, though not all of them near of the same Bigness, resembled one another in their shape, which was regular enough, and a very pretty one. But of this more elsewhere. §. I remember I have also long since taken pleasure to dissolve two or more of those saline Bodies, whose shapes we know already, in fair Water, that by a very gentle Evaporation I might obtain Concretions, whose Shapes should be, though curious, yet differing from the Figure of either of the Ingredients. But we must not expect, that, in all cases, the Salts dissolv'd together should be *totally* com-

compounded: for oftentimes they are of such different Natures, that one will shoot much sooner then another, and then it frequently happens, that a good Proportion of that will be first Chriftalliz'd in its own shape: as is confpicuoufly to be obferv'd in the refining of that impure Petre, (which, from the Country that affords it, the Purifiers call *Barbary* Nitre,) from the common Salt it abounds with: and (alfo) as *Agricola* obferves,* that in fome cafes, where a Vitriolate Matter is mingled with that, which yields Allom, thofe two kinds of Salts will shoot feparately in the fame large veffel, (which the Tryals, I have made with the compounded Solutions of thofe two Salts, do not difcountenance.) Now in fuch cafes, all that can be expected, or needs be defir'd, is, that the remaining part of the mixture, or fome portion of it, afford Chriftals, or Grains of compounded folid figures.

* *G. Agricola de re Metallica. lib. 11.*

R 4　　　Though

Though the *Venetian* Borax, wont to be sold in shops, be known to be a factitious Body, compounded of several Salts, that I shall not now stay to enumerate; and though, when we buy it, we usually find it to consist of Lumps and Grains mishapen enough, yet when I dissolv'd some of it in a good quantity of fair water, and made it coagulate very leisurely, I had Chrystals, upon whose surfaces I could perceive very exquisite and, as to sense, regular Geometrical figures. And one thing I must not here by any means prætermit, which is, that though the *Caput mortuum* of common *Aqua fortis* consists of Bodies of very differing Natures, (for such are Nitre and Vitriol,) and has been expos'd to a great violence of the Fire, yet *I* have sometimes admir'd the curiousness of those figures, that might be obtain'd barely by frequent Solutions and Coagulations of the Saline Particles of this *Caput mortuum* in fair water. But because

caufe the Glaſſes, wherein my Concretions were made, were too little to afford great Chriſtals, and they ought to
ſhoot very ſlowly; I choofe rather to
ſhew the Curious ſome large Chriſtals,
which I took out of the Laboratory of
an Ingenious Perſon, who, without
minding the Figures, had upon my Re-
commendation made great quantity of
that Salt, in large veſſels, for a Medicine:
(it being the *Panacea duplicata*, ſo famous in *Holſtein*.) For *divers* of theſe
Chriſtals have not onely Triangles,
Hexagons, and Rhomboids, and other
Figures exquiſitely Cut on their ſmooth
& ſpecular ſurfaces; and *others*, Bodies of
Priſmatical ſhapes: But ſome of them are
no leſs accurately figur'd then the fineſt
Nitre or Vitriol I remember my ſelf to
have obſerv'd, and ſome alſo terminate
in Bodies almoſt like Pyramids, conſiſting of divers Triangles, that meet in
one Vertical point, and are no leſs admi-
rably ſhap'd then the fairer ſort of
Corniſh

Cornish Diamonds, that have been brought me for Rarities. Befides, the producing of Salts of new fhapes, by compounding of Saline Bodies, I have found it to be practicable not onely in fome Grofs, or, as they fpeak, Corporal Salts, fuch as Sea-falt, Salt-petre, but alfo in fome Natural and fome Chymical Salts diffolv'd together; and, which perhaps you will think more confiderable in faline Spirits, made by diftillation: Not that all of them are fit for this purpofe, but that I have found divers of thofe, that work upon one another with Ebullition, to be fo. For in that Conflict the Saline Corpufcles come to be affociated to one another, and thereby, or by their newly acquir'd figure, whilft their Coalition lafts, to loofe much of their former Volatility: fo that, upon Evaporation of the fuperfluous Liquor, they will not fly, as otherwife they might; but concoagulate into finely fhap'd Chriftals, as I have try'd among

among other Saline Liquors, with Spirit of Urine, and Spirit of Nitre, and with Oyl of Vitriol, and Spirit of fermented Urine with Spirit of Sheeps bloud, and fpirit of Salt, and alfo with the Spirits of Salt and of Urine; which Laſt Experiment I the rather mention, becaufe it ſhews, by the difference of the Chriſtals, afforded by thofe two Liquors, from the Chriſtals refulting from one of them, namely the fpirit of Urine, (or if you pleafe, the Volatile Salt, wherewith it abounds,) concoagulated with a fit Dofe of Oyl of Vitriol, how much thofe compounded emergent figures depend upon the more fimple figures of the faline Corpufcles, that happen to convene into thofe new Concretes. For the fpirit of Urine, fatiated with fpirit of Salt, and both very gently, and not too far, Evaporated, often afforded me Chriſtals, that differ'd exceedingly in ſhape from thofe, which I obtain'd from the fame fpirit of Urine,

fatiated,

satiated, either with Oyl of Vitriol, or
with spirit of Nitre. For, (to adde
That upon the By,)that Salt,compoun-
ded of the two Spirits of Urine and of
common Salt, is wont to be very prettily figur'd, consisting of one long
Beam as it were, whence on both sides
issue out far shorter Christals , sometimes perpendicular to that, and paral-
lel to one another like the Teeth in a
Combe, and sometimes so inclining, as
to make the Whole appear almost like
a Feather;which is the more remarkable,
because I have (many years ago) observ'd, that common *Sal* Armoniack,
that is made of Urine and common Salt,
both crude, with a Proportion of Soot,
will, if warily dissolv'd, and coagulated,
shoot into Christals of the like shape.
How far the unknown Figure of a Salt
may Possibly (for I fear it will not Easily) be ghess'd at, by that of the Figure,
which it makes with some other Salt,
whose Figure is already known, I leave
to

to Geometricians to confider; having, I fear, infifted too long on this fubject already! But yet I muft adde one particular more, which will, as well illuftrate and confirme much of what has been faid above touching the Origination of Vitriol, as fhew, that the Shape of Vitriol depends upon the Textures of the Bodies, whereof it is compof'd.

Fourthly then, when I confider'd, that (as I formerly noted) Vitriol being but a Magiftery, made by the concoagulation of the Corpufcles of a diffolv'd Metal, with thofe of the Menftruum, the Magifteries of other Metals might, without inconvenience, be added, as other Vitriolate Concretes to the green, the blew, and white Vitriol, that are without fcruple referr'd to the fame *species*: and when I confider'd, that Oyl of Vitriol was not a fit Menftruum to *diffolve* divers of the Metals, nor even all thofe, that it will *corrode*; and that the like unfitnefs alfo is to be found in

com-

common spirit of Salt, I pitch'd upon *Aqua fortis* or spirit of Nitre, as that Menstruum, which was likeliest to afford variety of Vitriols: and accordingly I found, that besides the Lovely Vitriol of Copper formerly mention'd, that Liquor would with Quickfilver afford one fort of Chriftals, with Silver another, and with Lead a third; all which Chriftals of Vitriol, as they differ'd from each other in other Qualities, (upon which fcore you will find this Experiment elfewhere mention'd,) fo they did very manifeftly and confiderably differ in Shape: the Chriftals of Silver fhooting in exceeding thin Plates, and thofe of Lead and Quick-filver obtaining figures, though differing enough from each other, yet of a far greater depth and thicknefs, and leffe remote from the figure of common Vitriol or Sea falt: and yet all thefe Vitriols, efpecially That of crude Lead, when it was happily made, had Shapes curious and elaborate,

as

as well as thofe, we admire in common
Vitriol or Sea-falt.

IF then thefe Curious fhapes, which are
believed to be of the admirableft Effects, and of the ftrongeft Proofs of
fubftantial Forms, may be the Refults
of Texture, and if Art can produce Vitriol its felf, as well as Nature; why may
we not think, that in ordinary *Phænomena*, that have much lefs of wonder, recourfe is wont to be had to fubftantial
Forms without any Neceffity? (Matter,
and a Convention of Accidents being
able to ferve the turn without them;)
and why fhould we wilfully exclude
thofe Productions of the Fire, wherein
the Chymift is but a Servant to Nature,
from the number of Natural Bodies?
And indeed, fince there is no certain
Diagnoftick agreed on, whereby to difcriminate Natural and Factitious Bodies, and conftitute the *fpecies* of both;
I fee not, why we may not draw Arguments

ments from the Qualities and Operations of several of those, that are call'd
Factitious, to shew how much may be
ascrib'd to, and perform'd by, the Mechanical Characterization or *Stamp* of
Matter: Of which we have a noble Instance in Gunpowder, wherein by a bare
comminution and blending the Ingredients, Nitre, Charcoal, and Brimstone,
which have onely a new, and That an
exceeding slight Contexture, each retaining its own Nature in the Mixture;
so that there is no colour afforded to the
pretence of a substantial Form, there is
produc'd a new Body, whose Operations are more powerful and prodigious,
then those of almost any Body of Natures own compounding. And though
Glass be but an Artificial Concrete,
yet, besides that 'tis a very noble and
useful one, Nature her self has produc'd
very few, if enough, to make up a Number more lasting and more unalterable.
And indeed divers of those factitious
Bodies,

Bodies that Chymiſtry is able to afford
us, are endow'd with more various and
more noble Qualities, then many of
thoſe, that are unqueſtionably Natural.
And if we admit theſe Produ&ions into
the number of Natural Bodies, they will
afford us a multitude of Inſtances, to
ſhew, that Bodies may acquire many
and Noble Qualities, barely by having
Mechanical Affe&ions, introduc'd by
outward Agents into the Matter, or deſtroy'd there. As though Glaſs be ſuch
a Noble Body, as we have lately taken
notice of, yet ſince tis Fuſibility, Tranſparency, and Brittleneſs, that are its
onely Conſtituent Attributes, we can
in leſs then an hour, (or, perhaps halfe
that time,) turn an Opacous Body into
Tranſparent Glaſs, without the Addition of any other Viſible Body, by a
change of Texture, made in the ſame
Matter, and by another change of Texture, made without Addition, as formerly, we can, in a trice, reduce Glaſs into, or

S   obtain

obtain from it a Body, not Glaffy, but
Opacous, and otherwife of a very differing Nature, as it had been before. And
here let me adde what may not a little
conduce to our prefent Defign, That
even thofe, that imbrace *Ariftotle's*
principles, do unawares confeffe, that a
flight change of Texture, without the
introduction of a fubftantial Form, may
not onely make a Specifical difference
betwixt Bodies, but fo vaft a one, that
they fhall have differing *Genus'es*, and
may (as the Chymifts fpeak) belong to
differing Kingdoms. For Coral, to
pafs by all other Plants of that kind,
that may be mention'd to the fame purpofe, whilft it grows in the Bottom of
the Sea, is a real Plant, and feveral times
(which fuffices for my prefent fcope)
hath been there found by an Acquain-
tance of mine, as well as by other In-
quirers, foft and tender like another
Plant. Nay, I elfewhere* bring ve-

* In the Effays about things fuppofed to be fpontane-
oufly generated.

ry

ry good and recent Authority to prove,
that it is oftentimes found very fuccu-
lent, and does propagate its *fpecies*, as
well as other Shrubs; and yet Coral,
being gather'd and remov'd into the
Air, by the recefs of its Soul , no new
Lapidifick Form being fo much as pretended to, turns into a Concretion, that
is, by many Eminent Writers and others,
reckon'd among Lapideous ones: as indeed Coral does not burn like Wood,
nor obey Diftillation like it; and not
onely its *Calx* is very differing from the
Afhes of Vegetables, and is totally foluble in divers acid Liquors, and even
Spirit of Vinegar, but the uncalcin'd
Coral its felf will be eafily corroded
by good Vinegar, after the fame manner as I have feen *Lapis ftellaris*, and o-
ther unqueftionably Mineral ftones diffolv'd, fome by that Liquor, and fome
by the Spirit of it. A much ftranger
thing may be feen in the Eaft-India
Ifland of *Sombrero*, not very far from

Sumatra,

Sumatra, if we may believe our Coun-
tryman Sr *James Lancester*, who relates
it as an Eye witnesse, for which reason,
and for the strangenes of the thing, I
shall adde the story in his own words.
Here (*sayes he, speaking of the Coast
of Sombrero) *we found upon the sand by
the Sea side, a small Twigge growing up
to a young Tree, and offering to pluck up
the same, it shrunk down into the ground,
& sinketh, unless you hold very hard. And
being pluck'd up, a great Worme is the Root
of it: and look how the Tree groweth in
greatnes, the Worme diminisheth Now as
soon as the Worm is wholly turn'd into
the Tree, it rooteth in the ground, and
so groweth to be great. This Transformation was one of the greatest wonders
I saw in all my Travels. This Tree
being plucked up a little, the Leaves strip-
ped off and the Pill, by that time it was
dry turned into a hard Stone, much like
to white Coral. So that* (concludes he)

---

* *Purchas. Pilgr. part. the first. p. 152.*

this worme was twice transsformed into different natures: of these we gather'd and brought home many. The Industrious *Piso,* in his Excellent History of *Brasil,* vouches a multitude of Witnesses (not having Opportunity to be one himself) for the ordinary Transformation of a sort of Animals not much unlike Grass-hoppers) into Vegetables, at a certain season of the *year.

But since I sate down this Relation of S.r *John Lancester,* I have met with another, whose strangeness may much countenance it, in a small Tract newly publish'd by a Jesuite, *F. Michael Boym,* whom a good Critick much commended to me. For this Author doth, as an Eyewitnesse, affirme that, which is little lesse to my present Purpose. * *It vis,*

*i.e. I saw in a small fresh water, and shal-
low Lake of the Island* Hainan., (which
belongs to *China*) *Crabs, or Crawfishes,
which, as soon as they were drawn out of
the water, did in a moment loose both Life
and Motion, and became petrify'd, though
nothing appear'd to be chang'd either in
the External or Internal figure of their
Bodies.* What he further addes of thefe
Fifhes, is but of their Virtues in Phyfick, which, not concerning our fubject,
I fhall (*Pyrophilus*) willingly prætermit
it; and even, as to our Country-man's
relation, hoping, by means of an Ingenious Correfpondent in the Eaft-In-
dies, to receive a further Information a-
bout the ftrange Plant he mentions, I
fhall, at prefent, urge onely what has
been taken notice of concerning Coral, to countenance the Obfervation,
for whofe fake thefe Narratives have
been alleadg'd. And fo likewife, as to
what I was faying of Glafs, and Gunpowder, our receiving of thofe and the
gene-

generality of Factitious Bodies into the
Catalogue of Natural Bodies, is not
(which I formerly alfo intimated) neceffary to my prefent Argument: whereto it
is fufficient, that Vitriol is granted on all
hands to be a Natural Body, though
it be alfo producible by Art. And alfo
to the Argument it affords us, we might
adde that memorable Experiment deliver'd by *Helmont*, of turning Oyl of
Vitriol into Allom, by the Odour (as
he calls it) of Mercury, if, however it be
not defpicable, we had found it fit to be
rely'd on. But referving an Account
of that for another place, we fhall fubftitute the Inftance, prefented us by our
Author, about the Production of Salt-
petre: for if, having diffolv'd Pot-afhes
in fair water, you coagulate the filtrated Solution into a white Salt, and on
that pour Spirit of Nitre, till they will
not hifs any longer together, there will
fhoot, when the fuperfluous water is
Evaporated, Chriftals, that proclaim
S 4   their

their Nitrous Nature by their Prifma-
tical, (or at leaft Prifme-like) Shape,
their eafie Fufion, their Accenfion, and
Deflagration, and other Qualities, *partly* mention'd by our Author, and *partly*
difcoverable by a little Curiofity in
making Tryals.

II.

Experimental Attempts about the Redintegration of Bodies.

THe former of thofe two Arguments, (*Pyrophilus*) by which I
propof'd to confirme the Origine of
Forms, was, as you may remember,
grounded upon the Manner, by which
fuch a Convention of Accidents, as deferves to paffe for a Form, may be produc'd: and That having been hitherto
pro-

profecuted, it now remains, that we proceed to the Second Argument, drawn, not (as the former) from the firft Production, but from the Reproduction of a Phyfical Body. And though both thefe Arguments are valid; yet if this Latter could, in fpight of the Difficulties intervening in making of the Experiments that belong to it, be as clearly made out as the former, you would, I fuppofe, like it much the better of the two. For if we could Reproduce a Body, which has been depriv'd of its fubftantial Form, you would, I prefume, think it highly probable, if not more then probable, that (to borrow our Author's Expreffion) That which is commonly call'd the Form of a Concrete, which gives it its Being and Denomination, and from whence all its Qualities are in the Vulgar Philofophy, by I know not what inexplicable waies, fuppof'd to flow; may be in fome Bodies but a Characterization or Modifi.

cation

cation of the Matter they confift of; whofe parts, by being fo and fo difpof'd in relation to each other, conftitute fuch a determinate kind of Body, endow'd with fuch and fuch Properties, whereas, if the fame parts were otherwife dif- pofd, they wonld conftitute other Bo- dies, of very differing Natures from that of the Concrete, whofe parts they formerly were, and which may again refult or be produc'd, after its diffipati- on, and feeming deftruction, by the Re- union of the fame component Particles, affociated according to their former Di- fpofition.

But though it were not Impoffible to make an adæquate Redintegration of a Chymically Analiz'd Body, becaufe fome of the diffipated parts will either efcape through the Junctures of the Veffels, (though diligently clofd,) or, if they be very fubtle, will fly away up- on the difjoyning of the Veffels; or, will irrecoverably ftick to the infide of them:

them: yet I see not, why such a Reproduction, as is very possible to be effected, may not suffice to manifest what we intend to make out by it. For, even in such Experiments, it appears, that when the Form of a Natural Body is abolish'd, and its parts violently scatter'd; by the bare Reunion of some parts after the former manner, the very same Matter, the destroy'd Body was before made of, may, without Addition of other Bodies, be brought again to constitute a Body of the *like Nature* with the former, though not of *equal Bulk*. And indeed, the Experiment, recorded by our Author, about the Reproduction of Salt Petre, as it is the best and successfullest I have ever been able to make upon Bodies, that require a strong Heat to dissipate them; so I hope it will suffice to give you those thoughts about this matter, that the Author design'd in alledging it; and therefore, though having premis'd thus much, I shall proceed to

acquaint

acquaint you with the ſucceſs of ſome
Attempts he intimates (in that Eſſay)
his Intention of making, for the Redintegration of ſome Bodies; yet doing it
onely out of ſome Hiſtorical Notes I
find among my looſe Papers, That,
which I at preſent pretend to, is, but
partly to ſhew you the *difficulty* of ſuch
Attempts, which, ſince our Author's
Eſſay was communicated, have been
repreſented (I fear by Conjecture onely) as *very eaſie* to be accurately enough
done; and *partly*, becauſe our Author
does not, without reaſon, intimate the
uſefulneſs of Redintegrations, in caſe
they can be effected, and does, not cauſe-
leſly, intimate, that ſuch Attempts,
though they ſhould not Perfectly ſuc-
ceed, may increaſe the Number of Noble and Active Bodies, and conſequently, the Inventory of Mankind's Goods.
  Upon ſuch Conſiderations we attempted the Diſſipation and Reunion of the
parts of common Amber; and though
                         Chymiſts,

Chymifts, for fear of breaking their
Veffels, are wont, when they commit
it to diftillation, to adde to it a *caput
mortuum* (as they fpeak) of Sand, Brick,
&c. (in whofe room we fometimes
choofe to fubftitute beaten Glafs;)which
hinders them to judge of and employ
the Remanence of the Amber, after the
Diftillation is finifh'd: yet we fuppof'd,
and found, that if the Retort were not
too much fill'd, and if the Fire were
flowly and warily enough admin.fter'd,
the Addition of any other Body would
be needlefs. Wherefore having put
into a Glafs Retort four or five Ounces
of Amber, and admmiftred a gentle and
gradual heat, we obferv'd the Amber to
melt and bubble, (which we therefore
mention, becaufe ingenious men have
lately queftioned, whether it can be melted,) and having ended the Operation, &
fever'd the veffels, we found, that there
was come over in the form, partly of Oyl,
partly of Spirit & Flegme, and partly of
volatile

volatile Salt, near half the weight of the
Concrete: and having broken the Retort, we found, in the bottom of it, a
Cake of coal-black Matter, then whose
upper surface I scarce remember to have
seen in my whole life any thing more exquisitely polish'd; in so much, that, not-
withstanding the Colour, as long as I
kept it, it was fit to serve for a Looking Glass: and this smooth Mass being
broken, (for it was exceeding brittle,)
the larger fragments of it appear'd a-
dorn'd with an excellent lustre. All those
parts of the Amber, being put together
into a Glass Body, with a blind head luted to it, were placed in Sand, to be in-
corporated by a gentle heat: but whilst
I stept aside to receive a Visit, the Fire
having been increas'd without my
knowledge, the Fumes ascended so co-
piously, that they lifted up the Vessel
out of the Sand, whereupon falling a-
gainst the side of the Furnace, it broke
at the top, but, being seasonably call'd,
we

we fav'd all but the Fumes; and the remaining Matter looks not unlike Tarre, and with the leaft heat may be powr'd out like a Liquor, fticking even when it is cold to the fingers. Yet this open'd Body doth not eafily communicate fo much as a Tincture to fpirit of Wine, (which therefore feems fomewhat ftrange, becaufe another time prefumeing, that this would be a good way to obtain a Solution of fome of the refinous parts of Amber, we did, by pouring fpirit of Wine, that (though rectify'd) was not of the very beft, upon the reunited parts of Amber, lightly digefted into a Mafs, eafily obtain a clear Yellow Solution, very differing from the Tincture of Amber, and abounding (as I found by Tryal) in the diffolv'd fubftance of the Amber:) but in Oyl of Turpentine we have, in a fhort time, diffolv'd it into a bloud red Balfome, which may be of good ufe (at leaft) to Chirurgions. And having agen made the

former

former Experiment with more warmefs
then before, we had the like fuccefs in
our Diftillation, bur, the reunited parts
of the Amber being fet to digeft in a
large Boit head, the Liquor that was
drawn off, did, in a few hours, from its
own *Caput mortuum* extract a bloud red
Tincture, or elfe made a Solution of
fome part of it, whereby it obtain'd a
very deep Red, but having been, by intervening Accidents, hindred from fini-
fhing the Experiment, we mift the Satisfaction of knowing to what it may be
brought at laft.

And as for what our Author tels us
of this defign to attempt the Redintegration of Vitriol, Turpentine, and fome
other Concretes, wherein it feem'd not
unpracticable, he found in it more difficulty then every one would expect.
For the Bodies, on which fuch Experiments are likelieft to fucceed, feem to
be Allom, Sea falt, and Vitriol. And as
for Allom, he found it a troublefome
work

work to take (as a Spagirist would
speak) the Principles of it asunder, in
regard, that it is inconvenient to distill it
with a *Caput mortuum*, (as Chymists call
any fix'd Additament,) least that should
hinder the desir'd Redintegration of the
dissipated parts: And when he distill'd it
by its self, without any such Additament, he found, that, with a moderate
heat, the Allom would scarce part with
any thing but its Phlegm, and if he
urg'd it with a strong fire, he found, it
would so swell, as to endanger the breaking of the Retort, or threaten the boyling over into the Receiver. (Yet
having once been able very warily to
abstract as much Flegm and Spirit, as I
conveniently could, from a parcel of
Roch Allom, and having powr'd it back
upon that pulveriz'd *caput mortuum*, and
left the vessel long in a quiet place, I
found, that the Corpuscles of the Liquor, having had time, after a multitude
of Occursions, to accommodate and re-

T   unite

únite themselves to the more fix'd parts
of the Concrete, did by that Affociation (or Diffolution) recompofe, at the
top of the Powder, many Chriftalline
Grains of finely figur'd Salt, which increafing with time, made me hope, that, at
the length, the whole or the greateft part
would be reduc'd into Allom, which yet
a Mifchance, that robb'd me of the Glafs,
hindred me to fee.) So likewife of Sea
falt, if it be diftill'd, as it is ufual,
with thrice its weight of burn'd Clay, or
beaten Brick, twill prove inconvenient
in reference to its Redintegration; and
if it be diftill'd alone, it is apt to be fluxt
by the heat of the fire, and, whilft it remains in Fufion, will fcarce yield any
Spirit at all. And as for Vitriol, though
the Redintegration of it might feem to
be lefs hopeful, then that of the other
Salts, in regard that it confifts not onely of a Saline, but of a Metalline Body,
whence it may be fuppof'd to be of a
more intricate and elaborate Texture:
yet

yet becaufe there needs no *caput mortu-nm* in the Diftillation of it, we did, to purfue our Author's intimated defigns, make two or three Attempts upon it, and feem'd to mifs of our Aime, rather upon the Account of accidental hinde-rances, then of any infuperable difficul-ty in the thing it felf. For once, we with a ftrong fire, drew off from a parcel of common blew Vitriol, the Phlegm and Spirit, and fome quantity of the heavy Oyl, (as Chymifts abufively call it:) Thefe Liquors, as they came over with-out Separation, we divided into feveral parts, and the remaining very red *Caput mortnum* into as many. One of thefe parcels of Liquor we poured over night upon its correfpondent portion of the newly mentioned red Powder. But having left it in a Window, and the Night proving very bitter, in the mor-ning I found the Glafs crack'd in many places by the violence of the Froft, and the Liquor feem'd to have been funk'd

T 2

up

up by the Powder, and to have very much swelled it. This mixture then I took out, and placing it in an open mouth'd Glass in a Window, I found, after a while, divers Grains of pure Vitriol upon the other Matter, and some little Swellings, not unlike those we shall presently have Occasion to speak of. I took likewise a much larger parcel of the forementioned Liquor, and its correspondent proportion of *Caput mortuum*; and having leisurely mixt them in a large Glass Bason, I obtain'd divers *Phænomena*, that belong not to this place, but may be met with, where they will more properly fall in. In this Bason (which I lay'd in the Window, and kept from Agitation,) I perceived, after a while, the Liquor to acquire a blewish Tincture, and after ten or twelve weeks, I found the mixture dry, (for, it seems, it was too much expofed to the Air:) but the Surface of it adorn'd in divers places with Grains of Vitriol very curiously

riouſly figur'd. And beſides theſe, there
were ſtore of Protuberances, which conſiſted of aboundance of ſmall vitriolate
particles, which ſeem'd in the way to a
Coalition; for having let the Baſon a-
lone for four or five months longer, the
Matter appear'd cruſted over, partly
with very elevated Saline protuberances, partly with leſſer parcels, and partly
alſo with conſiderably broad Cakes of
Vitriol, ſome of above half an Inch in
breadth, and proportionably long; and
indeed the whole ſurface was ſo odly
diverſifi'd, that I cannot count the trouble, theſe Tryals have put me to, miſ-
pent.   Another time in a more ſlender
and narrow mouth'd Glaſs I pour'd back
upon the *Caput mortuum* of Vitriol the
Liquors, I had by violence of the fire
forc'd from it; ſo that the Liquid part
did ſwim a pretty height above the red
Calx, and remain'd a while limpid and
colourleſs: but the veſſel having ſtood,
for ſome time, unſtop'd in a Window,

the

the Liquor after a while, acquir'd by degrees a very deep vitriolate colour, and not long after, there appear'd, at the bottom and on the top of the *Calx*, many fair and exquisitely figur'd Grains of Vitriol, which cover'd the surface of the *Calx*; and the longer the vessel continu'd in the Window, the deeper did this Change, made upon the upper part of the Powder, seem to penetrate: so that I began to hope, that, in process of time, almost (if not more then almost) the whole mixture would be reduc'd to perfect Vitriol. But an Accident robb'd me of my Glass, before I could see the utmost of the Event.

And, on this Occasion, I must not prætermit an odd Experiment I lately made, though I dare not undertake to make it again. I elsewhere relate, how I digested, for divers weeks, a Quantity of powder'd Antimony, with a greater weight by half of Oyl of Vitriol, and how, having at length committed this

mix-

mixture to Diftillation, and thereby obtained, befides a little Liquor, a pretty
quantity of combuftible Antimonial or
Antimonio-Virriolate Sulphur ; there
remained, in the bottom of the Retort, a
fomewhat light and very friable *Caput*
mortuum, all the upper part of which
was at leaft as white as common Wood-
afhes, and the reft look'd like a Cinder.
And now I muft tell you what became
of this *Caput mortuum*, whereof I there
make no further mention. We could
not well forefee what could be made of
it, but very probable it was, that it
would afford us fome new Difcovery,
by being expofed to the fire, in regard
of the copious Sulphur, whereof it
feem'd to have been deprived: provided
it were urg'd in clofe Veffels, where nothing could be loft. Whereupon committing it to a naked fire in a fmall glafs
Retort, well Coated, and accommodated with a Receiver, we kept it there
many hours, and at length fevering the

li   T 4   Veffels,

Veflels, we found (which need not be wonder'd at) no Antimonial Quick-filver, and much lefs of Sulphur fublim'd then we expected: wherefore greedily haftning to the *Caput mortuum*, we found it flux'd into a Mafs, covered with a thin Cake of Glafs, whole fragments being held againft the light, were not at all coloured, as Antimonial Glafs is wont to be, but were as colourleffe as common white Glafs. The Lump above mentioned being broken, was found, fomewhat to our wonder, to be perfect black Antimony, adorn'd with long fhining ftreaks, as common Antimony is wont to be: onely this Antimony feem'd to have been a little refin'd by the fequeftration of its unneceffary Sulphur; which Ingredient feems by this Experiment, as well as by fome other Obfervations of ours, to be more copious in fome particular Parcels of that Mineral, then is abfolutely requifite to the conftitution of Antimony. Though

in

in our cafe it *may* be fufpected, that the
reduction of part of the Mafs to a colourlefs Glafs, was an effect of the Abfence
of fo much of the Sulphur, and might
in part make the remaining Maffe fome
amends for it.   What we further did
with this new or reproduced Concrete,
is not proper to be here told you: onely,
for your fatisfaction, we have kept a
Lump of it, that you may, with us, take
notice of what fome Philofophers
would call the Mindfulnefs of Nature,
which, when a Body was deprived of a
not inconfiderable portion of its chiefe
Ingredient, and had all its other parts
diffipated, and fhuffled, and difcolour'd,
fo as not to be knowable, was able to
rally thofe fcatter'd and difguifed parts,
and Marfhal or difpofe them into a Body of the former Confiftence, Colour,
&c. though (which is not here to be o-
verlook'd) the Contexture of Antimony, by reafon of the copious fhining
Styria, that enoble the darker Body, be
                                        much

much more elaborate, and therefore
more uneasie to be restored, then that
of many other Concretes.

But among all my Tryals about the
Redintegration of Bodies, That which
seem'd to succeed best, was made upon
Turpentine: for having taken some
Ounces of this, very pure, and good,
and put it into a Glass Retort, I distill'd
so long with a very gentle fire, till I had
separated it into a good quantity of very
clear Liquor, and a *Caput mortuum* very
dry and brittle: then breaking the Retort, I powder'd the *Caput mortuum*,
which, when it was taken out, was exceeding sleek, and transparent enough,
and very Red; but being powder'd, appear'd of a pure Yellow colour. This
Powder I carefully mixt vvith the Liquor, that had been distill'd from it,
vvhich immediately dissolv'd part of it
into a deep red Balsam; but by further
Digestion in a large Glass exquisitely
stopt, that Colour began to grovv fain-
ter,

ter, though the remaining part of the Povvder, (except a very little proportionable to fo much of the Liquor, as may be fuppof'd to have been vvafted by Evaporation, and Transfufion out of one Veffel into another,) be perfectly diffolv'd, and fo well reunited to the more fugitive parts of the Concrete, that there is fcarce any, that by the fmell, or taft, or confiftence, vvould take it for other then good and laudable Turpentine.

THE

✿✿✿✿✿✿✿✿✿✿✿✿✿✿✿✿✿✿✿✿✿✿✿✿✿

The I. Section *of the* Historical
Part (containing *the* Observations,
and beginning at pag. 107.) *is mis-
plac'd, and ought to have come in here,
and have immediately preceded this* II. Section *containing the* Experiments.

✿✿✿✿✿✿✿✿✿✿✿✿✿✿✿✿✿✿✿✿✿✿✿✿✿

❖❖❖❖❖❖❖❖❖❖❖❖❖❖❖❖❖❖❖❖❖

Advertisement *about the ensuing*
II. SECTION.

THe *Author would not have the Reader think, that the*
following Experiments, are the sole ones that he could
have set down to the same purpose with them. For they
are not the only that he had actually laid aside for this
occasion, till judging the ensuing ones sufficient for his
present scope, he thought it fitter to reserve Others for
those Notes *about the Production of particular Qualities,*
to which they seem'd properly to belong. Perhaps also it
will be requisite for me (because some Readers may think
the Omission a little strange) to excuse my having left di-
vers particulars unmentioned in more then One of the en-
suing Experiments. And I confesse that I might easily e-
nough both have taken notice of more Circumstances in them,
and made far more Reflections on them, if I would have
expatiated on the several Experiments according to the Di-
rections deliver'd in other* Papers. But though there, where
'twas my Design to give imployment to the Curiosity and
Diligence of as many Votaries to Nature, as (for want of
better instructions)had a mind to be so set on work, it was
fit the proposed Method should be suitable; yet here, where
I deliver Experiments, not so much as parts of Natural
History, as instances to confirm the Hypotheses, *and Dis-*
courses they are annexed to; it seemed needlesse, and im-
proper, (if not impertinent,) to set down Circumstances,
Cautions, Inferences, Hints, Applications, and other Par-
ticulars, that had no tendency to the scope, for which the
Experiments were alledged.

---

* Containing some Advices and Directions for the writing
of an Experimental Natural History.

☞ *These two Leaves are to be placed imme-*
diately before the 271 *page.*

And as for the kind of Experiments, here made choice of, I have the less scrupled to pitch upon Chymical Experiments, rather then Others on this occasion; not onely because of those Advantages which I have ascrib'd to such Experiments in the latter part of the Preface * to my Specimens, but because I have been Encouraged by the success of the Attempt made in those Discourses. For as new as it was when I made it four or five years ago, and as unusual a Thing as it could seem to divers Atomists and Cartesians, That I should take upon me to Confirm and Illustrate the Notions of the Particularian Philosophy (if I may so call it) by the help of an Art, which many were pleas'd to think cultivated but by Illiterate Operators, or if hymsical Phanaticks in Philosophy, and useful onely to make Medicines, or Disguize Metals: yet these Endeavours of ours met with much lesse opposition, then new Attempts are most commonly fain to struggle with. And in so short a time I have had the happiness to engage both divers Chymists to learn and relish the Notions of the Corpuscular Philosophy, and divers eminent Embracers of That. to endeavour to illustrate and promote the New Philosophy, by addicting themselves to the Experiments, and perusing the Books of Chymists. And I acknowledge, it is not unwelcome to me to have been (in some little measure) instrumental to make the Corpuscularian Philosophy, assisted by Chymistry, preferred to that which has so long obtained in the Schools. For (not here to consider, which I elsewhere do, how great an Advantage That Philosophy hath of This, by having an advantage of it in point of clearness,) though divers Learned and worthy men, that knew no better Principles, have, in cultivating the Peripatetick Ones, abundantly exercised and displaid their own Wit: yet I fear they have very little,

*. The Preface, here mentioned, is that premised to the Tract Intituled—— Some Speciment of an Attempt to make Chymical Experiments useful to illustrate the Notions of the Corpuscular Philosophy.

if

if at all, improved their Readers Intellect, or enrich it
with any true or ufeful Knowledg of Nature; but have
rather taught him to Admire Their Subtlety, then Underftand Hers. For to afcribe all particular Phænomena, that
feem any thing Difficult, (for abundance are not thought
fo, that are fo,) to fubftantial Forms, and, but nominal-
ly underftood, Qualities, is fo general and cafie a way
of refolving Difficulties, that it allows Naturalifts, without Difparagement, to be very Carelefs and Lazy, if it do
not make them fo: as in effect we may fee, that in about
2000 years fince Ariftotles time, the Adorers of his Phyficks, at leaft by vertue of His peculiar Princ ples, feem
to have done little more more then Wrangle, without clearing up (that I know of) any myftery of Nature, or producing any ufeful or noble Experiments: whereas the Cul-
tivators of the Particularian Philofophy, being obliged by
the nature of their Hypothefis, and their way of Reafo-
ning, to give the particular Accounts and Explications of
particular Phænomena of Nature, are alfo obliged, not
onely to know the general Laws and Courfe of Nature, but
to enquire into the particular Structure of the Bodies they
are converfant with, as that wherein, for the moft part,
their Power of acting, and Difpofition to be acted on, does
depend. And in order to this, fuch Enqiries muft take
notice of Abundance of Minute Circumftances; and to a-
void miftaking the Caufes of fome of them, muft often Make
and Vary Experiments; by which means Nature comes to
be much more diligently and induftrioufly Studied, and innumerable Particulars are difcover'd and obferved, which
in the Lazy Ariftotelian way of Philofophizing would not
be Heeded. But to return to that Decad of Inftances, to
which thefe Advertifements are premifed; I hope I need
not make an Apology for making choice rather of Chymical
Experiments, then others, in the fecond and concluding
Section

Section of the Historical Part of the present Treatise. But though I prefer that Kind of Instances, yet I would not be thought to overvalue Them in their kind, or to deny, that some Artists may (for ought I know) be found, to whose Chymical Arcana, these Experiments may be little better then Trifles. Nor perhaps are these the considerablest, that I my self could easily have communicated; (though these themselves would not be now Divulged, if I would have been ruled by the Disswasions of such as would have nothing of Chymical made Common, which they think Considerable.) But things of greater Value in themselves, and of Noble Use in Physick, may be less Fit for our present purpose, (which is not to impart Medicinal, or Alchymistical Processes, but illustrate Philosophical Notions,) then such Experiments as these; which, besides that they containe Variety of Phænomena, do not (for the most part) require either much Time, or much Charge, or much Skill.

The II, Sect.

The II. SECTION,
containing the
EXPERIMENTS.

Experiment I.

TAke good and clear Oyl of Vitriol, and caſt into it a convenient quan-tity of good Camphire groſly beaten; let it float there a while, and, without the help of external heat, it will infenſſibly be reſolv'd into a Liquor, which, from time to time, as it comes to be produc'd, you may, by ſhaking the Glaſs, mingle with the Oyl of Vitriol, whereunto you may, by this means, impart firſt a fine Yellow, and then a colour, which though it be not a true Red, will be of kin to it, and ſo very deep, as to make the mix-ture almoſt quite Opacous. When all

the

the Camphire is perfectly diffolv'd by incorporating with the Menſtruum, if you hit upon good Ingredients, and upon a right Proportion, (for a ſlight Miſtake in either of them, may make this part of the Experiment miſcarry,) you may probably obtain ſuch a mixture, as I have more then once had, namely, ſuch a one, as not onely to me, whoſe ſenſe of Smelling is none of the Dulleſt, but alſo to others, that knew not of the Experiment, ſeem'd not at all to have an Odour of the Camphire. But if into this Liquor you pour a due quantity of fair Water, you will ſee (perhaps not without delight) that, in a trice, the Liquor will become pale, almoſt as at the firſt, and the Camphire, that lay conceal'd in the pores of the Menſtruum, will immediately diſcloſe it ſelf, and emerge, in its own nature and priſtine form of white floating and combuſtible Camphire, which will fill not the Viol onely, but the neighbouring

part

part of the Air with its strong and Dif-
fufive Odour.

Now the *Phænomena* of this Experiment may, befides the ufes we elſewhere
make of it, afford us feveral particulars
pertinent to our prefent purpofe.

I. For (firſt) we fee a lighter and confiftent Body brought, by a Comminution, into Particles of a certain figure, to
be kept fwimming, and mixed with a
Liquor, on which it floated before, and
which is, by great odds, heavier then it
felf: fo that as by the Solution of Gold
in *Aqua regis*, it appears; that the pon-
peroufeſt of Bodies, if it be reduc'd to
parts minute enough, may be kept from
finking in a Liquor much lighter then it
felf: So this Experiment of Ours mani-
feſts what I know not whether hitherto
Men have prov'd, That the Corpufcles
of Lighter Bodies may be kept from e-
merging to the Top of a much heavier
Liquor: which Inftance being added to
that of the Gold, may teach us, that,
vvhen

when Bodies are reduc'd to very minute
parts, we muſt as well conſider their particular Texture, as the receiv'd Rules of
the Hydroſtaticks, in determining whether they will ſink, or float. or ſwim.

II. This Experiment alſo ſhews,
that ſeveral Colours, and even a very
deep one, may ſoon be produc'd by a
White Body, and a clear Liquor, and
that without the intervention of fire, or
any external heat.

III. And that yet this Colour may,
almoſt in the twinckling of an Eye, be
deſtroy'd, and as it were annihilated, and
the Latitant Whiteneſs, as many would
call it, may be as ſuddainly reſtor'd by
the Addition of nothing but fair Water, vvhich has no Colour of its ovvn,
upon vvhoſe account it might be ſur-
miſ'd to be contrary to the periſhing
colour, or to heighten the other into a
Prædominancy: nor does the Water
take into its ſelf, either the Colour it
deſtroy'd, or That it reſtores. For

IV.

IV. The more then femi-opacity of the Solution of Camphire and Oyl of Vitriol does prefently vanifh; and that Menftruum, with the Water, make up (as foon as the Camphorate Corpufcles come to be a float) one tranfparent and colourlefs Liquor.

V. And tis worth noting, that upon the mixture of a Liquor, which makes the Fluid much Lighter, (for fo Water is in refpect of Vitriol,) a Body is made to emerge, that did not fo, when the Fluid was much heavier. This Experiment may ferve to countenance what we elfewhere argue againft the Schools, touching the Controverfie about Miftion. For whereas though fome of them diffent, yet moft of them maintain, that the Elements alwaies loofe their Forms in the mix'd Bodies they conftitute; and though if they had dexteroufly propof'd their Opinion, and limited their Affertions to fome cafes, perhaps the Doctrine might be tolerated: yet fince

V   they

they are wont to propofe it crudely and
univerfally, I cannot but take notice,
how little tis favour'd by this Experiment; wherein even a mix'd Body (for
fuch is Camphire) doth, in a further miftion, retain its Form and Nature, and
may be immediately fo divorced from
the Body, to which it was united, as to
turn, in a trice, to the manifeft Exercife
of its former Qualities. And this Experiment being the eafieft Inftance, I
have devif'd, of the prefervation of a
Body, when it feems to be deftroy'd, and
of the Recovery of a Body to its former
Conditions; I defire it may be taken
notice of, as an inftance I fhall after have
Occafion to have recourfe to, and make
ufe of.

VI. But the notableft thing in the
Experiment is, that Odours fhould depend fo much upon Texture; that one
of the fubtleft and ftrongeft fented
Drugs, that the Eaft it felf or indeed
the World affords us, fhould fo foon
quite

quite loofe its Odour, by being mix'd
with a Body that has fcarce, if at all, any
fenfible Odour of its own, and This,
while the Camphorate Corpufcles fur-
vive undeftroy'd, in a Liquor, from
whence one would think, that leffe fubtle and fugitive Bodies, then they, fhould
eafily exhale.

VII. Nor is it much leffe confiderable, that fo ftrong and piercing a Sent
as that of Camphire, fhould be, in a
moment, produc'd in a Mixture, wherein none of it could be perceiv'd before,
by fuch a Liquor as Water, that is quite
devoid of any Odour of its own: which
fo eafie and fuddain reftauration of the
Camphire to its Native Sent, as well as
other Qualities, by fo languid a Liquor
as common Water, doth likewife argue,
that the Union or Texture of the two
Ingredients, the Camphire and the Oyl
of Vitriol, was but very flight, upon
which nevertheleffe a great alteration in
point of Qualities depended. And to con-

firme

firme, that divers of the præceding *Phæ:*
nomena depend upon the particular Texture of the Liquors, imploy'd to exhibit
them, I fhall add, that if, inftead of oyl of
Vitriol, you caft the Concrete into well
deflegm'd Spirit of Nitre, you will obtain no red, nor dark, but a Tranfparent
and Colourlefs Solution. And when
to the above mention'd red Mixture I
put, inftead of fair Water, about 2 or 3
parts of duely rectifi'd Spirit of Wine,
there would enfue no fuch changes, as
thofe formerly recited; but the Spirit
of Wine, that diffolv'd the Concrete,
when it was by it felf, without loofing
its Diaphaneity, or acquiring any Colour, did, when it diffolv'd the Mixture, diffolve it with its new adventitious Colour, looking like a grofs red
Wine, fomewhat turbid, or not yet well
freed from its Lees: fo that this Colour
appear'd to refide in the Mixture as
fuch, fince neither of the two Ingredients diffolv'd in, or mingled vvith the
Spirit

Spirit of Wine, would have afforded
that Colour, or indeed any other. But if
to this Liquor, that look'd like troubled Wine, we poured a large Proportion of fair Water, the Rednefs would
immediately vanifh , and the Whole
would, *as to fenfe*, become White
throughout; I fay, *as to fenfe*, becaufe
the Whiteneffe did not indeed appertain properly to the whole Mixture,
but to a huge multitude of little Corpufcles of the reviv'd Concrete, whereof fome or other, which at firft fwamme
confufedly to and fro, left no fenfible
Portion of the Liquor unfurnifh'd with
fome of them; whereas when the Camphorate Corpufcles had leifure to e-
merge, as they foon did, they floated in
the forme of a White Powder or Froth
at the top of the Liquor, leaving all the
reft as clear and colourleffe as the common Water.

But we have not yet mention'd all
the ufe, we defign'd to make of our

Mixture, for by profecuting the Experiment a little further, we made it afford
us fome new *Phænomena*.

VIII. For having kept the Mixture
in a moderately warme place, (which
circumftance had perhaps no influence
on the Succeffe,) and having diftill'd it
out of a Glafs Retort, the Event anfwer'd our Expectation, and the Liquor,
that came over, had a Sent; which,
though very ftrong, was quite differing
both from that of the Mixture, and that
of the Camphire; and in the remaining
Body, though the Liquor and the Camphire it confifted of, were either both
tranfparent, or the one tranfparent as a
Liquor, and the other white, as tranfparent and colourleffe Bodies are wont
to be made by Contufion: yet the remaining Mafs , which amounted to a
good part of the Mixture, was not onely Opacous, but as black as Coal, in
fome places looking juft like polifhed
Jet;

Jet; which is the more coufiderable, becaufe that though Vegetable Subſtances, that are not fluid, are wont to acquire a Blackneſs from the fire, yet neither do Liquors, that have already been diſtill'd, obtain that Colour upon Rediſtillation; neither have we, upon Tryal purpoſely made, found, that Camphire, expoſ'd to fire in a Retort, fitted with a Receiver, (which was the caſe of the preſent Experiment,) would at all acquire a Jetty Colour, but would either totally aſcend White, or afford *Flores*, and a *Caput mortuum* (as a vulgar Chymiſt would call the Remaines) of the ſame Colour, both in reſpect of one another, and in reſpect of the Camphire.

IX. And our Experiment afforded this notable *Phænomenon*, That though Oyl of Vitriol be a diſtill'd Liquor, and though Camphire be ſo very fugitive a Subſtance, that being left in the Air, it will, of it ſelf, fly all away; and therefore

V 4   Phy-

Phyſicians and Druggiſts preſcribe the keeping it in Linſeeds Or *Millium*, or other convenient Bodies, to hinder its Avolation; yet, by our Experiment, its Fugacity is ſo reſtrain'd, that not onely the *Caput mortuum* newly mention'd, endured a good fire in the Retort, before it was reduc'd to that pitchy Subſtance vve vvere lately mentioning, but having taken ſome of that ſubſtance out of the Retort, & order'd it, by a carefulWorkman, to be kept in a cloſely cover'd Crucible during ſome time in the fire; when it vvas brought me back, after the Pot had been kept red hot above half an hour, there remain'd a good quantity of the Matter, brittle, vvithout any ſmell of Camphire, and as black as ordinary Charcoal; ſo much do the Fixity and Volatility of Bodies depend upon Texture.

Experiment.

Experiment II.

AMong thofe other Experiments of mine, *(Pyrophilus)* which tend to manifeft, that new Qualities may be produc'd in Bodies, as the Effects of new Textures; I remember, fome years ago, I writ for a Friend a whole Set of Tryals, that I had made about the Changes I could produce in Metals and Minerals, by the Intervention of Sublimate. But though the whole Tract, wherein they are recited, might be pertinent enough to our prefent Subject; yet referving other paffages of it for other places, (efpecially for our Notes upon thofe particular Qualities, which they are moft proper to illuftrate,) it may at this time fuffice me to fend you a Tranfcript of what that Account contains, relating to Copper and Silver, the one a mean and fugitive, and the other a noble and fix'd Metal. For thofe changes

in

Coloûr, Confiſtence, Fuſibleneſſe, and other Qualities, which you will meet with in theſe Experiments, will afford us divers *Phænomena*, to ſhew what great Changes may be made, even in Bodies ſcarce corruptible, by one or more of thoſe three Catholick wayes of Natures working according to the Corpuſcular Principles, namely, the Acceſs, the Receſs, and the Tranſpoſition of the minute Particles of Matter.

As for my Method of changing the Texture of Copper, I confeſs it hath oftentimes ſeem'd ſtrange to me, that Chymiſts, plainly ſeeing the notable Effect, that Sublimate, diſtill'd from Antimony, has upon that Mineral, by opening it, and volatilizing it, (as we ſee it do in the making of what they are pleaſ'd to call *Mercurius vitæ*,) ſhould not have the Curioſity to try, whether or no Sublimate might not likewiſe produce, if not the ſame, yet a conſiderable

rable Change in other Mineral Bodies, there appearing no reafon, or at leaft there having been none given, that I know of, why the Referating Operation (if I may fo fpeak) of Sublimate, fhould be confin'd to Antimony. Upon thefe Confiderations, we were invited to endeavour to fupply the Neglect we had obferv'd in Chymifts, of improving the Experiment of *Butyrum Antimonti*: and though an Indifpofition in point of Health, which befell us before we had made any great progrefs in our Enquiries, made us fo fhy of the Fumes of Sublimate and Minerals, that we neither did make all our Tryals fo accurately, nor profecute them fo far as we would have done, had we been to deal with more innocent Materials: Yet we fuppofe, it will not be unwelcome to You, to receive from us a naked, but faithful, Narrative of our Proceedings; being apt to think, that you will therein find Inducements to carry on this

Expe-

Experiment further then we have done, and to compleat what we have but begun.

Firſt then, we took half a pound of Copper plates, of about an Inch broad, and the thickneſs of a Grain of Wheat, (which we after found was too great,) and of an arbitrary length; then caſting a Pound of groſly beaten *Venetian* Sublimate into the bottom of a ſomewhat deep Glaſs Retort, we caſt in the Copper plates upon it, that the Fumes of the Sublimate might, in their Aſcenſion, be compell'd to act upon the incumbent Metal, and then placing this Retort, as deep as we well could, in a Sand Furnace, and adapting to it a ſmall Receiver, we adminiſter'd a Gradual fire ſeaven or eight hours, and at length for a while increaſ'd the heat, as much as we well could do in ſuch a Furnace. The ſucceſs of this Operation was as follows.

1. There came little or no Liquor at all over into the Receiver, but the Neck & upper

upper part of the Retort were Candied on the infide, by reafon of the copious Sublimate adhæring to them, which Sublimate weigh'd above Ten Ounces; in the Retort we found about two Oun- ces and a quarter of running Mercury, which had been fuffer'd to revive by the acid Salts, which corroding the Copper, forfook the Quickfilver, whereto they had been in the Sublimate united.

2. Upon the increafe of the fire, there was plainly heard a Noife, made by the melting Matter in the Retort, not un- like that of a boyling Pot, or of Vitri- ol, when being committed to a Calci- ning fire, it is firft brought to flow. And this Noife we found to be a more con- ftant Circumftance of this Experiment, then the revification of part of the Mer- cury contain'd in the Sublimate; for up- on another Tryal, made with the former proportion of Copper plates and Sub- limate, we obferv'd, during a very long while, fuch a Noife as hath been already

men-

mention'd, but the Operation being finish'd, we scarce found so much as a few Grains of running Mercury, either in the Retort or Receiver.

3. We found the Metalline Lump, in the bottom of the Retort, to have been increas'd in weight somewhat more then (though not half an Ounce above) two Ounces; some of the Copper plates, lying at the bottom of the Mass, retain'd yet their Figure and Malleableness, which we ascrib'd to their not having been thin enough to be sufficiently wrought upon by the Sublimate: the Others, which were much the greater number, had wholly lost their Metalline form, and were melted into a very brittle Lump, which I can compare to nothing more fitly, then a lump of good Benjamin; for this Mass, though ponderous, was no less brittle, and being broken, appear'd of divers Colours, which seem'd to be almost transparent, in some places it was red, in others of a high

high and pleafant Amber Coloür, and
in other parts of it, Colours more dar-
kifh and mix'd might be difcern'd.

4. But this ftrange Mafs being broken into fmaller Lumps, and laid upon
a Sheet of White Paper in a Window,
was, by the next morning, where ever
the Air came at it, all cover'd with a
lovely greenifh Blew, or rather, blewifh
Green, almoft like that of the beft Ver-
degreefe, and the longer it lay in the air,
the more of the internal parts of the
Fragments did pafs into the fame Colour: but the vvhite Paper, which in
fome places they ftain'd, feem Dy'd of
a Green colour inclining unto Yellow.
And here we had Occafion to take notice of the infinuating fubtlety of the
Air; for having put fome pieces of this
Cupreous Gum (if I may fo call it) into
a little Box, to fhut out the Air, which
vve have found it poffible to exclude
by other means, vve found, that notwithftanding our care, thofe included
Fragments

Fragments were, as well as the reft already mention'd, covered with the powder, as it were of *viride Æris*.

5. We muft not, on this Occafion, o-
mit to tell you, that, having, the laft year,
made fome Tryals in reference to this
Experiment, we obferv'd in one of
them, that fome little Copper plates,
from which Sublimate had been drawn
off, retain'd their priftine fhape, and
Metalline nature, but were Whitened
over like Silver, and continu'd fo for
divers Months, (though we cannot precifely tell you hovv long, having at
length accidentally loft them.) And to
try vvhether this Whitenefs vvere onely fuperficial, vve purpofely broke fome
of thefe flexible Plates, and found, that
this Silver colour had penetrated them
throughout, and vvas more glorious in
the very Body of the Metal, then on its
Surface, vvhich made us fufpect, that
the Sublimate, by us imploy'd, had been
adulterated vvith Arfenick, (vvhere-
vvith

with the Sophifticators of Metals are wont to make Blanchers for Copper, but not to mention, that rhe Malleable-neffe continu'd, which Arfenick is wont to deftroy,)we difcover'd not by Tryal, that the Sublimate was other then fin-cere.

6. In this Metalline Gum the Body of the Copper appear'd fo chang'd and open'd, that we were invited to look upon fuch a Change as no ignoble Ex-periment, confidering the Difficulty, which the beft Artifts tell us there is, and which thofe,that have attempted it, have found, I fay not, to unlock the Sulphur of Venus, but to effect leffe Changes in its Texture, then was here-by made. For this Gum, caft upon a quick Coal, and a little blown, will partly melt and flow like Rofin, and partly flame, and burn like a Sulphur, and with a flame fo lafting, if it be re-kindled as often as it leaves off burning, that we obferv'd it, not without fome

X   Wonder;

Wonde, and so inflammable is this ope-
ned Copper, that, being held to the
flame of a Candle, or a piece of lighted
Paper, it would almost in a moment
take fire, and send forth a flame like
common Sulphur, but onely that it
seem'd to us to incline much more to a
greenish colour, then the blewer flame
of Brimstone is wont to do.

To these *Phænomena* of our Experiment, as it was made with Copper, my
Notes inable me to subjoyn some o-
thers, exhibited when we made it with
Sublimate and Silver.

There were taken of the purest sort of
Coined Silver we could get, half a score
thin Plates, on which vvas cast double
the vveight of Sublimate in a small and
strongly coated Retort. This Matter
being sublim'd in a naked fire, vve
found, (having broken the Vessel,) that
the Sublimate vvas almost totally ascended to the top and neck of the Retort, in the latter of vvhich appear'd in

many

(293)

many places fome reviv'd Mercury, in
the bottom of the Retort we found a
little fluxed Lump of Matter, which
'twas fcarce poffible to feparate from
the Glafs, but having, with much adoe
divorc'd them, we found this Mafs to
be brittle, of a pale yellowifh colour, of
neer about the weight of the Metal, on
which the Sublimate had been caft. And
in the thicker part of this Lump there
appear'd, when it was broken, fome part
of the Silver plates, vvhich, though brittle, feem'd not to have been perfectly
diffolv'd. This Refin of Silver did, like
that of Copper, but more flowly, im-
bibe the Moifture of the Air, and vvith-
in about 24 hours, vvas cover'd vvith a
fomevvhat greenifh Duft, concerning
vvhich vve durft not determine, vvhether it proceeded from that mixture of
Copper, vvhich is generally to be met
vvith in coyned Silver, or from the
compounded Metal. Nor the more
curious fort of Painters do, as they in-

form

form us, by corroding coined Silver
vvith the fretting steams of saline Bodies, or vvith corrofive Bodies themfelves, turn it into a fine kind of Azure,
as we may elfevvhere have opportunity
more particularly to declare.   I fhall
novv onely adde, that fome fmall fragments of our Refin, being caft upon red
hot Coals, did there vvaft themfelves
in a flame not very differing in colour
from that of the former mention'd Refin of Copper, but much more durable
then vvould have eafily been expected
from fo fmall a quantity of Matter.

This is all the Account I can give you
of our firft Tryal, but fufpecting, that
the Copper, vvont to be mixt as an
Alloy vvith our coyned Silver, might
have too much Influence on the recited
Event; coming aftervvards into a place,
vvhere vve could procure Refin'd Silver, vve took an Ounce of That, and
having Laminated it, vve caft it upon
tvvice its Weight of beaten Sublimate,
                                    vvhich

which being driven away from it with a
fomewhat ftrong fire, we took, out of
the bottom of the Glafs Retort, a Lump
of Matter, which in fome places, where
it lay next the Glaffe, was as it were
filver'd over very finely, but fo very
thinly, that the Thickneffe of the Silver
fcarce equall'd that of fine white Paper;
the reft of the Metal (except a little
that lay undiflolv'd almoft in the middle of the Maffe, becaufe, as we fuppof'd, the Plates had not been beaten,
till they were fufficiently and equally
thin,) having been, by the faline part of
the Sublimate, that ftuck to it, colliquated into a Mafs, that look'd not at all like
Silver, or fo much as any other Metal
or Mineral.

And tis remarkable, that though Silver be a fixt Metal, and accounted indeftructible, yet it fhould by fo flight an
Operation, and by but about a quarter
of its vveight of Additament, (as appear'd by weighing the whole Lump,)

X 3   be

be fo ftrangely difguized, and have its Qualities fo alter'd.

For (firft) though an eminent White-neffe be accounted the colour, which belongs to pure Silver, and though beaten Sublimate be alfo eminently White, yet the Mafs, we are fpeaking of, was partly of a Lemmon or Amber colour, or a deep Amethyftinine colour, aud partly of fo dark a one, as it feem'd black: and it was pretty, that fometimes in a fragment, that feem'd to be one continued and entire piece, the upper part would be of a light Yellow, vvhich abruptly ending, the lower vvas of a colour fo obfcure, as fcarce to challenge any name diftinct from Black.

Next whereas Silver is one of the moft Opacous Bodies in Nature, and Sublimate a White one, the produc'd Mafs was in great part Tranfparent, though not like Glafs, yet like good Amber.

Thirdly, the Texture of the Silver
vvas

was exceedingly alter'd: for our Mass;
instead of being Malleable and Flexible,
as that Metal is very much, appear'd, if
you went about to cut it with a Knife,
like Horn, yet otherwise easily apt to
crack and break, though not at all to
bend.

Fourthly, whereas Silver will indure
Ignition for a good while before it be
brought to Fusion, our Mixture will
easily melt, not onely upon quick coals,
but in the flame of a Candle; but this
Resin, or Gum (if I may so call it) of
our fix'd Metal did not, like that, we formerly describ'd, of Copper, tinge the
flame of a Candle, or produce with the
glowing coals, on which 'tis laid, either a
green or blewish colour.

And (*Pyrophilus*) to discover how
much these Operations of the Sublimate upon Copper and Silver depend
upon the particular Textures of these
Bodies, I took two parcels of Gold, the
one common Gold thinly laminated,

and

and the other very well refin'd, and having caſt each of theſe in a diſtinĉt Urinal, upon no leſs then thrice its weight of groſly beaten Sublimate, I cauſ'd this laſt nam'd ſubſtance to be, in a Sand furnace, elevated from the Gold, but found not, that either of the two Parcels of that Metal was manifeſtly alter'd thereby: whether in caſe the Gold had been reduc'd to very minute particles, ſome kind of change (perhaps, if any differing enough from thoſe lately recited to have been made in the Copper and the Silver) might have been made in it, I am not ſo abſolutely certain; but I am confident, that by what I reſerve to tell you hereafter of Sublimates Operation upon ſome other Minerals, eſpecially Tin, it will appear, that That Operation depends very much upon the particular Texture of the Body, from whence that ſublimate is Elevated.

Before I diſmiſs this ſubjeĉt, *Pyrophilus*, I muſt not conceale from you, that

in

in the Papers, whence thefe Experiments made with Sublimate have been
tranfcribed, I annex'd to the whole Difcourfe a few Advertifements, whereof
the firft was, That I was reduc'd, in thofe
Experiments, to imploy, for want of a
better, a Sand Furnace, wherein I could
not give fo ftrong a fire as I defir'd,
which circumftance may have had fome
Influence upon the recited *Phænomena*;
and among other Advertifements there
being one, that will not be impertinent
to my prefent Defign, and may poffi-
bly afford a not unfuccefsful Hint, I fhall
fubjoin it in the words, wherein I find it
deliver'd.

The next thing, of which I am to ad-
vertife you, is this, That this Experiment may probably be further im-
prov'd, by imploying about it various
and new kinds of Sublimate, and that
feveral other things may be fublim'd up
together either with crude Mercury, or
with common Sublimate, he that confiders

fiders the way of making vulgar Sublimate, will not, I fuppofe, deny. To give
you onely one Inftance, I fhall inform
you, that, having caufed about equal parts
of common Sublimate and Sal Armoniack to be well powder'd and incorpora-
ted, by fubliming the Mixture in ftrong
and large Urinals plac'd in a Sand
Furnace, we obtain'd a new kind of Sublimate, differing from the former, which
we manifefted *ad Oculum*, by diffolving
a little of it and a little of common Sublimate feverally in fair water; for dropping a little refolv'd falt of Tartar upon
the folution of common Sublimate, it
immediately turn'd of an Orange tawny
colour, but dropping the fame Liquor
upon the folution of the Ammoniack
Sublimate, if I may fo call it, it prefently turn'd into a Liquor in Whiteneffe
refembling Milk: And having from 4
ounces of Copper plates drawn 6 ounces
of this new Sublimate after the already
often recited manner, we had indeed in
the

the bottom of the Retort a **Cupreous**
Refin; not much unlike That, made by
Copper and common Sublimate; and
this Refin did, like the other, in the
moift Air, foon begin to degenerate into
a kind of Verdigreefe. But that which
was fingular in this Operation was, that
not onely fome of the Sublimate had
carried up, to a good height, enough of
the Copper to be manifeftly colour'd
by it of a fine blewifh Green, but into
the Receiver there was pafs'd neer an
Ounce of Liquor, that fmelt almoft like
fpirit of Sal Armoniack, and was tincted
like the Sublimate, fo that we fuppof'd
the Body of the Venus to have been
better wrought upon by this, then by
the former Sublimate. And yet I judg'd
not this way to be the moft effectual
way of improving common Sublimate,
being apt to think, upon grounds not
now to be mention'd, that it may, by
convenient Liquors, be fo far enrich'd
and advanc'd, as to be made capable of

opening

opening the Compact Body of Gold it felf, and of producing in it fuch Changes, (which yet perhaps will enrich but mens Underftandings,) as Chymifts are wont very fruitlefly to attempt to make in that almoft Indeftructible Metal. But of This, having now given you a Hint, I dare here fay no more.

Experiment III.

THere is (*Pyrophilus*) another Experiment , which many wil find more eafie to be put in practice, and which yet may, as to Silver, be made a kind of *Succedaneum* to the former, and confequently may ferve to fhew, how the like Qualities in Bodies may be effected by differing Wayes, provided a like Change of Texture be produc'd by them. Of This I fhall give you an Example in that Preparation of Silver , that fome Chymifts have call'd *Luna Cornea*, which I fhall not fcruple to

men.

mention particularly, and apply to my
present purpofe; becaufe though the
name of *Luna Cornea* be already to be
met with in the Writings of fome Alchymifts, yet the thing it felf, being not
uf'd in Phyfick, is not wont to be known
by thofe that learn Chymiftry in order
to Phyfick; and the way that I ufe in
making it is differing from that of Alchymifts, being purpofely defign'd to
fhew fome notable *Phænomena*, not to
be met with in their way of proceeding.

We take then refined Silver, and having beaten it into thin Plates, and diffolv'd it in about twice its Weight of
good *Aqua fortis*, we Filtrate it carefully to obtain a clear folution, (which
fometimes we Evaporate further, till it
fhoot into Chryftals, which we afterwards dry upon brown Paper with a
moderate heat.)

Upon the abovemention'd folution
we drop good fpirit of Salt, till we find,
that it will no more curdle the Liquor
it

it falls into, (which will not happen fo
foon, as you will be apt at firft to imagine,) then we put the whole Mixture
in a Glafs Funnel lin'd with Cap-paper,
and letting the moifture drain through,
we dry, with a gentle heat, the fubftance,
that remains in the Filtre, firft wafhing
it (if need be) from the loofly adhæring
Salts, by letting fair Water run through
it feveral times, whilft it yet continues
in the Filtre. This fubftance being
well dry'd, we put it into a Glafs Viol,
which being put upon quick coals, firft
cover'd with Afhes, and then freed from
them, we melt the contain'd fubftance
into a Mafs, which, being kept a while in
Fufion, gives us the *Luna Cornea* we
are now to confider.

If to make this Factitious Concrete,
we firft reduce the Silver into Chryftals,
and afterwards proceed with fpirit of
Salt, as we have juft now taught you to
do with the folution, we have the exceedingly Opacous, Malleable, and hard-

ly

ly Fusible Body of Silver, by the convenient interposition of some saline Particles, not amounting to the third part of the Weight of the Metal, reduc'd into Chryftals, that both shoot in a peculiar and determinate figure, differing from those of other Metals, and also are diaphanous and brittle, and by great odds more easily fusible then Silver it self; besides other Qualities, wherein having elsewhere taken notice, that these Chryftals differ both from Silver and from *Aqua fortis*, we shall not now insist on them, but pass to the Qualities, that do more properly belong to the change of the Solution of Silver into *Luna Cornea*.

First then we may observe, that though spirit of Salt be an highly acid Liquor, and though acid Liquors and *Alkalys* are wont to have quite contrary Operations, the one præcipitating what the other would diffolve, & diffolving what the other would præcipitate: yet in our
cafe,

cafe, as neither Oyl of Tartar *per deliquium*, nor fpirit of Salt will diffolve
Silver, fo both the one and the other
will præcipitate it; which I defire may
be taken notice of againft the Doctrine
of the Vulgar Chymifts, and as a Proof,
that the Præcipitation of Bodies depends not upon acid or Alkalizate Liquors as fuch, but upon the Texture of
the Bodies, that happen to be confoun-
ded.

2. We may here obferve, that Whitenefs and Opacity may be immediately
produc'd by Liquors, both of them
Diaphanous and colourlefs.

3. That on the other fide, a White
Powder, though its minute parts appear not tranfparent, like thofe of beaten Glafs, Rofin, &c. which, by comminution, are made to feem White, may
yet, by a gentle heat, be prefently rednc'd
into a Mafs indifferently Tranfparent,
and not at all White, but of a fair Yellow.

4. We

4. We may obferve too, that though Silver require fo ftrong a fire to melt it, and may be long kept red hot, without being brought to Fufion; yet by the affociation of fome faline particles, conveniently mingled with it, it may be made fo fufible, as to be eafily and quickly melted, either in a thin Viol, or at the flame of a Candle, where it will flow almoft like Wax.

5. It may alfo be noted, that though the Lunar folution and the fpirit of Salt would, either of them apart, have readily diffolv'd in Water; yet when they are mingled, they do, for the moft part, concoagulate into a fubftance, that will lie undiffolv'd in Water, and is fcarce, if at all, foluble either in *Aqua fortis*, or in fpirit of Salt.

6. And remarkable it is, that the Body of Silver being very flexible and malleable, (efpecially if the Metal be, as ours was, refin'd) it fhould yet, by the Addition of fo fmall a proportion of

Y   falt,

Salt, (a Body rigid and brittle,) as is affociated to it in our Experiment, be
made of a Texture fo differing from
what either of its Ingredients was before, being wholly unlike either a Salt
or a Metal, and very like in Texture to
a piece of Horn. And to fatisfie my
felf, how much the Toughnefs of this
Metalline Horn depended upon the
Texture of the *Compofitum*, refulting
from the refpective Textures of the feveral Ingredients, I præcipitated a folution of Silver with the diftill'd faline Liquor commonly call'd Oyl of Vitriol,
inftead of fpirit of falt, and having wafh'd
the Præcipitate with common Water, I
found agreeably to my conjecture, that
this Præcipitate, being flux'd in a moderate heat, afforded a Mafs, that look'd
like enough to the Concrete we have
been difcourfing of , but had not its
Toughnefs, being brittle enough to be
eafily broken in pieces. But the two
confiderableft *Phænomena* of our Experiment

Experiment do yet remain unmentiond.
For 7ᵗʰˡʸ. Tis odd, that whereas a
folution of Silver is, as we have often
occafion to note, the bittereft Liquor
we have ever met with, and the fpirit of
Salt far fowrer then either the fharpeft
Vinegar, or even the fpirit of it, thefe
two fo ftrongly and offenfively tafted
Liquors fhould be fo eafily and fpeedily, without any other thing to correct
them, be reduc'd into an infipid fubftance, (at leaft fo far infipid, that I have
lick'd it feveral times with my Tongue,
without finding it otherwife, though
perhaps, with much rowling it to and fro
in the mouth, it may at length afford
fome unpleafant Taft, but exceedingly
different from that of either of the Liquors ¡that compof'd it:) and This,
though the Salts, that made both the
Silver, and the præcipitating fpirit fo
ftrongly tafted, remaine affociated with
the Silver.

8. And Laftly, it is very ftrange,
that

that though the faline Corpufcles, that give the efficacy both to good *Aqua fortis*, and the like fpirit of Salt, be not onely fo volatile, that they will eafily be diftill'd with a moderate fire, but fo fugitive, that they will in part fly away of themfelves in the cold Air, (as our Nofes can witnefs to our trouble, when the Viols, that contain fuch Liquors, are unftopt;) yet by vertue of the new Texture they acquire, by affociating themfelves with the Corpufcles of the Silver and with one another, thefe minute particles of falt loofe fo much of their former Lightnefs, and acquire fuch a degree of Fixednefle, that they will endure melting with the Metal they adhere to, rather then fuffer themfelves to be driven away from it. Nor do I remember, that when I melted this Mafs in a thin Viol, I could perceive any fenfible Evaporation of the Matter: nay having afterwards put a parcel of it upon a quick Coal, though that were blown

to

to intend the heat; yet it ſuffer'd Fuſion, and ſo ran off from the Coal, without appearing, when it was taken up a-
gain, to be other then *Luna Cornea,* as it
was before.

Experiment IV.

I Am now (*Pyrophilus*) about to do a
Thing, contrary enough both to my
Cuſtome and Inclination, that is, To
diſcourſe upon the *Phænomena* of an
Experiment, which I do not teach you
to make. But ſince I cannot as yet,
without ſome breach of promiſe, plainly
diſcloſe to you what I muſt now conceal, your Equity aſſures me of your
Pardon. And as, becauſe the Qualities
of the Salt, I am to ſpeak of, are very
remarkable, and pertinent to my preſent
deſign, I am unwilling to paſs them by
unmention'd; ſo I hope, that notwith-
ſtanding their being ſtrange, I may be
allow'd to diſcourſe upon them to you,
Y 3   vvho,

who, I prefume, know me too well to fufpect I would impofe upon you in matters of fact, and to whom I am willing (if you defire it) to fhew the Anomalous Salt it felf, and Ocular proofs of the chief properties I afcribe to it.

I fhall not then fcruple to tell you, that Difcourfing one day with a very Ingenious Traveller and Chymift, who had had extraordinary Opportunities to acquire Secrets, of a certain odd Salt I had thought upon and made, which was of fo differing a kind from other Salts, that though I did not yet know what Feats I fhould be able to do with it, yet I was confident, it muft have Noble and unufual Operations. This Gentleman, to requite my Francknefs, told me, that I had lighted on a greater Jewel, then perhaps I was aware of; and that if I would follow his Advice, by adding fomething that he nam'd to me, and profecuting the Preparation a little further, I fhould obtain a Salt exceeding-

ly

ly noble. I thank'd him, as I had caufe, for his Advice, and, when I had Opportunity, follow'd it. And though I found the vvay of making this Salt fo nice and intricate a thing, that if I vvould, I could fcarce eafily defcribe it, fo as to enable moft men to practice it; yet having once made it, I found, that, befides fome of the things I had been told it would perform, I could do divers other things vvith it, vvhich i had good caufe to believe the Gentleman, of whom I was fpeaking, did not think of; and I doubt not, but I fhould have done much more with it, if I had not unfortunately loft it foon after I had prepar'd it.

Several of the *Phænomena*, I try'd to produce with it, which are not fo proper for this place, are referv'd for another, but here I fhall mention a few, that beft fit my prefent purpofe.

Firft then, though the feveral Ingredients, that compof'd this Salt, were all of them fuch, as Vulgar Chymifts muft

Y 4   according

according to their Principles, look upon as purely Saline, and were each o:
them far more falt then Brine, or mor:
fowr then the ftrongeft Vinegar, or more
ftrongly tafted then either of thofe two
Liquors; yet the Compound, made up
of onely fuch Bodies, is fo far from being eminently falt, or fowr, or infipid,
that a Stranger being ask'd, what Taft it
had, vvould not fcruple to judge it rather fweet, then of any other Taft:
though its Sweetnefs be of a peculiar
kind, as there is a difference even among
Bodies fweet by Nature; the fweetnefs
of Sugar being divers from that of Ho-
ney, and both of them differing from
that of the fweet Vitriol of Lead. And
this is the onely inftance, I remember, I
have hitherto met vvith of Salts, that,
vvithout the mixture of infipid Bodies,
compofe a fubftance *really fweet*. I fay
really fweet, becaufe Chymifts oftentimes terme the *Calces* of Metals and
other Bodies *dulcifi'd*, if they be freed
from

from all corrofive falts and fharpnefs of
Taft, *fweet*, though they have nothing
at all of pofitive fweetnefs in them; and
by that licence of fpeaking do often e-
nough impofe upon the Unskilful.

Another thing confiderable in our A-
nomalous Salt is, That though its O-
dour be not either ftrong or offenfive,
(both which that of Volatile Salts is
wont to be,) yet if it be a little urg'd
with heat, fo as to be forc'd to evaporate haftily and copioufly, I have known
fome, that have been uf'd to the powerful ftink of *Aqua fortis*, diftill'd Urine,
and even fpirit of Sal Armoniack its
felf, that have complain'd of this fmell,
as more ftrong, and upon that account
more unfupportable then thefe themfelves: and yet when thefe Fumes fettle
again into a Salt, their Odour will again
prove mild and inoffenfive, if not pleafant.

Thirdly, whereas all the Volatile, and
Acid, and Lixiviate Salts, that we know
of,

of, are of fo determinate and fpecificated
a Nature, (if I may fo fpeak,) that there
is no one fort of the three, but may be
deftroy'd by fome one or other of the
other two Salts, if not by both, as fpirit
of Urine, which is a volatile Salt, being
mingled with fpirit of Salt, or *Aqua fortis*, or almoft any other ftrong and acid
fpirit, will make a great Ebullition, and
loofe its peculiar Taft, and feveral of its
other Qualities; and on the otherfide,
Salt of Tartar, and other *Alkalys*, (that
is, Salts produc'd by Incineration of
mix'd Bodies,) will be deftroy'd with
Ebullition by *Aqua fortis*, fpirit of Salt,
or almoft any other ftrong fpirit of that
Family. And fpirit of Salt, *Aqua fortis*, &c. will be (as they fpeak) deftroy'd
both by Animal volatile Salts, and by
the fix'd Salts of Vegetables; that is,
will make an Effervefcence with either
fort of Salts, and compofe with them a
new Liquor or Salt, differing from either
of the ingredients, and, as to taft, fmell,
odour,

odour, and divers other Qualities, more languid and degenerous: whereas, I fay, each of thefe three Families of Salts may be eafily deftroy'd by the other two, our Anomalous Salt feems to be above the being *thus wrought* upon by a-ny of all the three, and is the onely Body I know: (which is no fmall priviledge, or rather prerogative,) for I did not find, that a Solution of it, made with as little Water as I could, which is the vvay whereby we ufually make it fluid, would make any Ebullition, either with Oyl of Tartar *per Deliqnium*, or fpirit of Sal Armoniack, or ftrong fpirit of Salt, or even Oyl of Vitriol, but would calmely and filently mix vvith thefe differing Liquors, and continue as long as I had patience to look upon them, without being præcipitated by them. But this is not the onely way I imploy'd to examine, whether our Salt belong'd to any of the three above mention'd comprehenfive families of Salts. For

I found not, that the strongest solution
of it would turn Syrup of Violets either
red, as acid spirits do, or green, as both
fix'd and volatile Salts will do.   Nor
would our Solution turn a clear one of
Sublimate made in common Water, either white, as spirit of Urine, Sal Armoniack, or others of the same family,
or into an Orange Tawny, like salt of
Tartar, and other *Alkalys*: but left the
solution of Sublimate transparent, without giving it any of these colours, mingling it self very kindly with it, as it had
done with the four lately mention'd
Liquors. And to satisfy my self a little
further, I not onely try'd, that an undis-
colour'd mixture of syrup of Violets
and our solution, would immediately be
turn'd red by 2 or 3 drops of spirit of
Salt, or green by as much Oyl of Tartar: but, to prosecute the Experiment,
I let fall a drop or two of a mixture
made of our Anomalous solution, and
spirit of Salt well shaken together, upon
some

some syrup of Violets, which was thereby immediately turn'd red, and a little
of the same Anomalous solution, being
shaken together with Oyl of Tartar *per
Deliquium*, turn'd another parcel of the
same syrup of Violets into a delightful
green, which, hapning as *I* expected,
seem'd to argue, that our Solution,
though as to sense it were exquisitely
mingled in the several mixtures, to
which *I* had put it, did, as it left them
their undestroy'd respective Natures,
retain its own; and yet this Salt is so far
from being a languid or an insignificant
thing, that *Aqua fortis*,and Oyl of Vitriol themselves, as operative and as fu-
rious Liquors as they are, are unable in
divers cases to make such Solutions,and
perform such other things,as our calme,
but powerful, Menstruum can, though
but slowly,effect.

Fourthly: Though this Salt be a volatile one, and requires no strong heat
to make it sublime into finely figur'd
Chry-

Chryſtals without a remanence at the Bottom; yet being diſſolv'd in Liquors, you may make the Solution, if need be, to boile, without making any of the Salt ſublime up, before the Liquor be totally or almoſt totally drawn off, whereas the volatile ſalt of Urine, Bloud, Harts-horn, &c. are wont to aſcend before almoſt any part of the Liquor, they are diſſolv'd in, which is in many caſes very inconvenient.

And though this be a Volatile ſalt, yet I remember not, that _I_ have obſerv'd any fix'd ſalt, (without excepting ſalt of Tartar it ſelf,) that runs near ſo ſoon _per Deliquium,_ as this will do; but by abſtraction of the adventitious moiſture tis eaſily reſtor'd to its former ſaline form: and yet differs from ſalt of Tartar, not onely in Fixedneſſe and Taſt, and divers other qualities, but alſo in this, That, whereas ſalt of Tartar requires a vehement fire to flux it, a gentlier heat, then one would eaſily imagine,

gine, will melt our Salt into a Limpid
Liquor.

And whereas fpirit of Wine will diffolve fome Bodies, as Sanderick, Ma-
ftick, Gum-Lac, &c. and Water, on the
other fide, diffolves many that fpirit of
Wine cannot, and Oyls will diffolve
fome, for which neither of the other
Liquors are good folvents; our falt will
readily diffolve both in fair Water, in
the higheft rectifi'd fpirit of Wine, (and
That fo little, as not to weigh more
then the falt,) and in Chymical Oyls
themfelves, with which it will affociate
its felf very ftrictly, and perhaps more
too, then *I* have yet found any other
confiftent falt to do.

Experiment V.

THe Experiment *I* am *(Pyrophilus)*
now about to deliver, though I
have not yet had Opportunity to perfect what *I* defign'd, when fome Notions

ōns, that I have about Fire and Salt, fug-
gefted it to me, is yet fuch as may far
more clearly, then almoft any of the Experiments commonly known to Chymifts, ferve to fhew us, how near to a
real Tranfmutation thofe Changes may
prove, that may be effected even in inanimate, and, which is more, fcarce corruptible Bodies, by the recefs of fome
Particles, and the accefs of fome others,
and the new Texture of the refidue.
The Experiment I have made feveral
wayes, but one of the lateft and beft I
have uf'd is this: Take one part of good
Sea-falt well dry'd and powder'd, and
put to it double its weight of good *A-*
qua fortis, or fpirit of Nitre, then have-
ing kept it (if you have time) for fome
while in a previous digeftion, diftill it
over with a flow fire in a Retort or a
low Body, till the the remaining Matter be quite dry, and no more; for this
fubftance, that will remain in the bottom of the Glafs, is the thing that is
fought for. This

This Operation being performable in a moderate fire, and the Bodies them-felves being almoft of an incorruptible nature, one would fcarce think, that fo flight a matter fhould produce any Change in them; but yet I found, as I expected, thefe notable Mutations of Qualities effected by fo unpromifing a way.

For in the firft place, we may take notice, that the Liquor, that came over, was no longer an *Aqua fortis*, or fpirit of Nitre, but an *Aqua Regis*, that was able to diffolve Gold, which *Aqua fortis* will not meddle with, and will not dif-folve Silver, as it would have done be-fore, but will rather, as I have purpofe-ly try'd, præcipitate it out of *Aqua for-tis*, if that Menftruum have already dif-folv'd it: but this Change belonging not fo properly to the fubftance it felf I was about to confider, I fhall not here infift on it.

2. Then, the Taft of this Subftance

Z   comes

comes by this Operation to be very
much alter'd. For it hath not that ſtrong
ſaltneſs that it had before, but taſts far
milder, and, though it rellish of both,
affects the Palate much more like Salt-
petre, then like common ſalt.

3. Next, whereas this laſt nam'd Body is of very difficult Fuſion, our factitious ſalt imitates ſalt-petre in being very fuſible, and it will, like Nitre, ſoon
melt, by being held in the flame of a
Candle.

4. But to proceed to a more conſiderable *Phænomenon*, tis known, that
Sea-ſalt is a Body, that doth very much
reſiſt the fire, when once by being
brought to Fuſion, it hath been forc'd
to let go that windy ſubſtance, that
makes unbeaten ſalt crackle in the fire,
and ſo by blowing it accidentally increaſe it. Tis alſo known, that acid ſpi-
rits, as thoſe of Salt, Vitriol, Nitre, Vinegar, &c. are not onely not inflammable themſelves, but hinderers of inflam-
mation

mation in other Bodies; and yet my
Conjecture leading me to expect, that,
by this Operation, I should be able to
produce, out of two inflammable Bodies, a third, that would be easily inflammable. I found, upon Tryal, not onely that small Lumps of this substance,
cast upon quick and well blown coals,
though they did not give so blew a
flame as Nitre, did yet, like it, burn a-
way with a copious and vehement flame.
And, for further Tryal, having melted
a pretty quantity of this transmuted
Sea salt in a Crucible, by casting upon
It little fragments of well kindled Charcoal, it would, like Nitre, presently be
kindled, and afford a flame so vehement
and so dazling, that one that had better
Eyes then I, and knew not what it was,
complain'd, that he was not able to support the splendor of it. Nor were all its
inflammable parts consum'd at one deflagration: for by casting in more fragments of well kindled Coal, the Matter

would

would fall a puffing, and flame afresh for several times consecutively, according to the quantity that had been put into the Crucible.

5. But this it self was not the chief discovery I design'd by this Experiment. For I pretended hereby to devise a way of turning an acid salt into an *Alkaly*, which seems to be one of the greatest and difficultest Changes, that is rationally to be attempted among durable and inanimate Bodies. For 'tis not unknown to such Chymists as are any thing inquisitive and heedful, how vast a difference there is between acid Salts, and those, that are made by the combustion of Bodies, and are sometimes call'd Fix'd, sometimes Alkalizate. For whereas strong Lixiviums (which are but strong solutions of *Alkalys*) will readily enough dissolve common Sulphur, and divers other Bodies abounding with Sulphur; even those highly acid Liquors, *Aqua fortis*, and *Aqua Regis*,

Regis, though fo corrofive, that one will diffolve Silver, and the other Gold it felf, will let Brimftone lye in them undiffolv'd I know not how long; though fome fay, that in procefs of time, there may be fome Tincture drawn by the Menftruum from it, which yet I have not feen try'd; and though it were true, would yet fufficiently argue a great difparity betwixt thofe acid fpirits, and ftrong Alkalizate folutions, which will fpeedily diffolve the very maffe of common Sulphur. Befides, 'tis obferv'd by the inquifitive Chymifts, nor does my Experience contradict it, that the Bodies, that are diffolv'd by an acid Menftruum, may be præcipitated by an Alkalizate; and on the contrary, folutions, made by the latter, may be præcipitated by the former. Moreover, as Litharge, diffolv'd in fpirit of Vinegar, will be præcipitated by the Oyl of Tartar *per Deliquium*, or the folution of its Salt; and, on the contrary, Sulphur or

Z 3   Anti-

Antimony, diſſolv'd in ſuch a ſolution, will be præcipitated out of it by the ſpirit of Vinegar, or even common Vinegar. Moreover, Acids and Alkalizates do alſo differ exceedingly in taſt, and in this greater diſparity, that the one is volatile, and the other fix'd, beſides other particulars not neceſſary here to be inſiſted on. And indeed, if that were true, which is taught in the Schools, that there is a natural enmity, as well as diſparity betwixt ſome Bodies, as between Oyly and wateriſh ones, the Chymiſts may very ſpecioullly teach, (as ſome of them do) That there is a ſtrange contrariety betwixt Acid and Alkalizate Salts; as when there is made an Affuſion of oyl of Tartar upon *Aqua Regis*, or *Aqua fortis*, to præcipitate Gold out of the one, and Silver out of the other, their mutual Hoſtility ſeems manifeſtly to ſhew it ſelf, not onely by the noiſe, and heat, and fume, that are immediately excited by their

conflict,

conflict, but by this moſt of all, that afterwards the two contending Bodies will appear to have mutually deſtroy'd one another, both the ſowr Spirit and the fixt Salt having each loſt its former Nature in the ſcuffle, and degenerated with its Adverſary into a certain Third ſubſtance, that wants ſeveral of the Properties both of the ſowr Spirit and the *Alkaly*. Now to apply all this to the Occaſion, on which I mention'd it, how diſtant and contrary ſoever the more inquiſitive of the latter Chymiſts take *Acid* and *Fixed* Salts to be; yet I ſcarce doubted, but that, by our Experiment, I ſhould, from acid ſalts, obtain an *Alkaly*, and accordingly having, by caſting in ſeveral bits of well kindled coal, excited, in the melted Maſs of our tranſmuted Salt, as many Deflagrations as I could, and then giving it a pretty ſtrong fire to drive away the reſt of the more fugitive parts, I judg'd, that the remaining Maſſe would be (like the fix'd

Nitre

Nitre I have elfewhere mention'd) of an Alkalizate nature, and accordingly having taken it out, I found it to taft, not like Sea-falt, but fiery enough upon the Tongue, and to have a Lixiviate relifh. I found too, that it would turn Syrup of Violets into a greenifh colour, that it would præcipitate a Limpid folution of Sublimate, made in fair water, into an Orange tawny Powder. I found, that it would, like other fix'd falts, produce an Ebullition with acid fpirits, and even with fpirit of falt it felf, and concoagulate with them. Nor are thefe themfelves all the wayes I took to manifeft the Alkalizate Nature of our tranfmuted Sea falt.

I did indeed confider at firft, that it might be fufpeﬅed, that this new Alkalizateneffe might proceed from the Afhes of the injeﬅed Coals, the Afhes of Vegetables generally containing in them more or leffe of a fix'd Salt. But when I confider'd too, that a pound of
Char-

Charcoal, burn'd to Afhes, is wont to yield fo very little Salt, that the inje-cted fragments of Coal, (though they had been, which they were not) quite burn'd out in this Operation, would fcarce have afforded two or three grains of falt, (perhaps not half fo much,)I faw no reafon at all to believe, that in the whole Mafs I had obtain'd (and which was all, that was left me of the Sea-falt, I had at firft imploy'd,) it was nothing but fo inconfiderable a proportion of Afhes, that exhibited all the *Phænomena* of an *Alkaly*.

And for further confirmation both of This, and what I faid a little before, I fhall adde, that to fatisfie my felf yet more, I pour'd, upon a pretty quantity of this Lixiviate falt, a due proportion of *Aqua fortis*, till the hiffing and ebullition ceafed, and then leaving the fluid Mixture for a good while to coagulate, (which it did very flowly,) I found it at length to fhoot into faline Chryftals,

ſtals, which though they were not of the figure of Nitre, did yet, by their in. flammability and their bigneſs, ſuffici- ently argue, that there had been a Con. junction made betwixt the Nitrous ſpirit, and a conſiderable proportion of *Alkaly.*

I conſider'd alſo, that it might be ſuſ- pected, that in our Experiment twas the Nitrous Corpuſcles of the *Aqua fortis,* that, lodging themſelves in the little rooms deſerted by the ſaline Corpuſ- cles of the Sea-ſalt, that paſs'd over into the Receiver, had afforded this *Alkaly*; as common Salt.petre, being handled after ſuch a manner, would leave in the Crucible a fix'd or *Alkalizate* Salt. But to this I anſwer, that as the Sea-ſalt, which was not driven over by ſo mild a Diſtillation, and ſeem'd much a greater part then that which had paſs'd over, was far from being of an *Alkalizate* na- ture: ſo the Nitrous Corpuſcles, that are preſum'd to have ſtay'd behind, were

whilſt

whilſt they compoſ'd the ſpirit of Niː
tre, of an highly volatile and acid Naː
ture, and conſequently of a nature direꞋ
ctly oppoſite to that of *Alkalys*; and if
by the addition of any other ſubſtance,
that were no more *Alkalizate* then Sea-
ſalt, an *Alkaly* could be obtain'd out of
ſpirit of Nitre or *Aqua fortis*, the Pro-
duciblenefs of an *Alkaly* out of Bodies
of another nature might be rightly
thence inferr'd: ſo that however, it appears, that by the intervention of our
Experiment, two Subſtances, that were
formerly acid, are turn'd into one, that is
manifeſtly of an *Alkalizate* Nature,
which is That we would here evince.

Perhaps it may (*Pyrophilus*) be worth
while to ſubjoyn; That to proſecute
the Experiment by inverting it, we
drew two parts of ſtrong ſpirit of Salt
from one of purifi'd Nitre; but did not
obſervc the remaining Body to be any
thing neer ſo conſiderably chang'd as
the Sea-ſalt, from which we had drawn
the

the fpirit of Nitre; fince though the fpirit of Salt, that came over, did (as we ex-
pe&ed) bring over fo many of the Corpufcles of the Nitre, that, being heated,
it would readily enough diffolve foliated
Gold; yet the Salt, that remain'd in the
Retort, being put upon quick Coals, did
flafh away with a vehement and halitu-
ous flame, very like that of common
Nitre.

Experiment VI.

I Come now (Pyrophilus) to an Experiment, which, though in fome
things it be of kin to that which I have
already taught you , concerning the
changing of Sea-falt by Aqua fortis, will
yet afford us divers other inftances, to
fhew, how upon the change of Texture
in Bodies , there may arife divers new
Qualities; efpecially of that fort, which,
becaufe they are chiefly produc'd by
Chymiftry, and are wont to be confider'd

der'd by Chymifts, if not by Them onely, may in fome fenfe be call'd Chymical.

The Body, which, partly whilft we
were preparing it, and partly when we
had prepar'd it, afforded us thefe various *Phænomena*, either is the fame that
Glauberus means by his *Sal Mirabilis*,
or at leaft feems to be very like it: and
whether it be the fame or no, its various
and uncommon Properties make it very
fit to have a place allow'd it in this
Treatife. Though of the many Tryals I
made with it, I can at prefent find no
more among my loofe Papers, then that
following part of it, that I wrot fome
years ago to an Ingenious Friend, who
I know will not be difpleaf'd, if, to fave
my felf fome time, and the trouble of
Examining my Memory, I annex the
following Tranfcript of it.

[To give you a more particular account of what I writ to you from *Oxford* of my Tryals about *Glauber's* Salt,
though

though I dare not fay, that I have made the felf fame Thing, which he cals his *Sal Mirabilis*, becaufe he has defcrib'd it fo darkly and ambiguoufly, that tis not eafie to know with any certainty what he means; yet whether or no I have not made Salt, that, as far as I have yet try'd it, agrees, well enough with what he delivers of His, and therefore is like to prove either his *Sal mirabilis*, or almoft as good a one, I fhall leave you to judge by this fhort Narrative.

The ftrange things that the Induftrious *Glauber*'s Writings have invited Men to expect from his *Sal mirabilis*, in cafe he be indeed poffeff'd of fuch a thing, and the Enquiries of divers Eminent Men, who would fain learn of me, what I thought of its Reality and Nature, invited me, the next Opportunity I got, to take into my hands his *Pars altera Miraculi Mundi*, whofe Title you know promifes a Defcription of this *Sal Artis mirificum*, as he is pleaf'd to call it. But,

I

I confeſs, I did not read it near all over, becauſe a great part of it is but a Tranſcription of ſeveral entire Chapters out of *Paracelſus*, and I perceiv'd, that much of the reſt did, according to the cuſtome or Chymical Writings, more concern the Author, then the ſubjeɛt; wherefore looking upon his proceſs of making his *ſal mirabilis*, I ſoon perceiv'd he had no mind to make it common, ſince he onely bids us upon two parts of common Salt diſſolv'd in common Water, to pour *A*, without telling us what that *A* is, wherefore reading on in the ſame proceſſe, and finding that he tels us, that with *B* (which he likewiſe explaines not at all, nor determines the quantity of it) one may make an *Aqua fortis*, it preſently call'd into my mind, That ſome Years before, having had Occaſion to make many Tryals, mention'd in other Traɛts of mine, with Oyl of Vitriol and Salt petre, I did, among other things, make a red ſpirit of Nitre, by the help

one-

onely of Oyl of Vitriol; remembring This (I fay) I reforted to one of my *Carneades's* Dialogues,* and reviewing that Experiment, as I have fet it down, I concluded, That though I had not diffolv'd the Salt petre in Water, as *Glauber* doth his common Salt; yet fince, on the other fide, I made ufe of external fire, 'twas probable I might this way alfo get a Nitrous fpirit, though not fo ftrong. And though by calling the Liquor, that muft make an *Aqua fortis B*, whereas he had call'd that, which is to make his fpirit of Salt and *fal mirabilis*, *A*, he feem'd plainly to make them differing things, yet relying on the Experiment I had made, and putting to a folution ot Nitre as much of the Oyl of Vitriol as I had taken laft, though That be double the quantity he prefcribes for the making of his *Sal mirabilis*, I obtain'd, out of a low glaffe Body and Head plac'd in Sand, an indif-

---

* See the Sceptical Chymift.

ferent

ferent good *Spiritus Nitri*, that even
before Rectification would readily e-
nough diffolve Silver, though it were
diluted with as much of the common
Water, wherein Salt-petre had been
diffolv'd, as amounted at leaft to double or treble the weight of the Nitrous
parts, the remaining Matter, being kept
in the fire till it was dry, afforded us a
Salt eafily reducible (by Solution in
fair Water and Coagulation)into Chryftalline Grains, of a nature very differing
both from crude Nitre, and from fixt
Nitre, and from Oyl of Vitriol. For it
coagulated into pretty big and well
fhap'd Grains, which, you know, fix'd
Nitre and other *Alkalizate* Salts are
not wont to do; and thefe Graines were
not like the Chryftals of Salt-petre it
felf, long and Hexaedrical, but of another figure, not eafie nor neceffary to be
here defcribed.

Befides, this Vitriolate Nitre (if I
may fo call it) would not eafily, if at all,

flow,

flow in the Air, as fixt Nitre is vvont
to do. Moreover, it was eafily enough
fufible by heat, vvhereas fix'd Nitre
doth ufually exact a vehement Fire for
its Fufion; and though crude Salt-petre
alfo melts eafily, yet to fatisfie you how
differing a fubftance this of ours was
from That, vve caft quick Coals into
the Crucible, without being at all able
to kindle it. Nay, and vvhen, for further Tryal, vve threw in fome Sulphur
alfo, though it did flame away it felf,
yet did it not feem to kindle the Salt,
that was hot enough to kindle It; much
lefs did it flafh, as Sulphur is wont on
fuch occafions to make Salt-petre do.
Add to all this, That a parcel of this
white fubftance, being, vvithout Brimftone, made to flow for a vvhile in a
Crucible, with a bit of Charcoal for it to
vvork upon, grew manifeftly and ftrongly fented of Sulphur, and acquir'd an
Alkalixate Taft, fo that it feem'd almoft a Coal of fire upon the Tongue,

it

if it were lick'd before it imbib'd any of
the Aires moifture, and (which many
perhaps will, though I do not, think
ftranger)obtain'd alfo a very red colour;
which recall'd to my mind, that *Glauber*
mentions fuch a Change obfervable in
his Salt, made of common Salt, upon
whofe Account he is pleaf'd to call fuch
a fubftance his *Carbunculus*.

Being invited by this fuccefs to try,
whether I could make his *Sal mirabilis*,
notwithftanding his intimating, as I
lately told you, that it is done with a
differing Menftruum from that, wherewith the Salt-petre is to be wrought
upon; I obferv'd, that where he points
at a way of making his Salt in quantity
without breaking the Veffels, he prefcribes, that the Materials be diftill'd in
Veffels of pure Silver; vvhence I conjectur'd, that 'twas not *Aqua fortis*, or
fpirit of Nitre, that he imploy'd to o-
pen his Sea-falt: and that confequently,
fince common fpirit of Salt was too

weak to effect fo great a Change, as the Experiment requires, 'twas very probable, that he imploy'd Oyl of Sulphur, or of Vitriol, vvhich vvill fcarce at all fret unalloy'd Silver. And however I concluded, that whatfoever the Event fhould prove, it could not but be worth the While to try, vvhat Operation fuch a Menftruum vvould have upon Sea-falt, as I vvas fure had fuch a notable one upon falt-petre. And I remember, that formerly making fome Experiments about the differing manners of Diffolution of the fame Concrete by feveral Liquors, I found, that Oyl of Vitriol diffolves Sea.falt in a very odd way, (vvhich you vvill find mention'd among my promifcuous Experiments,) vvherefore pouring, upon a folution of Bay-falt, made in but a moderate proportion of Water, Oyl of Vitriol to the full Weight of the dry Salt, and abftracting the Liquor in a Glafs Cucurbite plac'd in Sand, I obtain'd, without
ftrefs

ſtreſs of fire, beſides flegme, good ſtore
of a Liquor, vvhich, by the Smel and
Taſt, ſeem'd to be ſpirit of Salt. And
to ſatisfie my ſelf the better, mingling
a little of it vvith ſome of the ſpirit of
Nitre lately mention'd, I found the mixture, even without the Aſſiſtance of
Heat, to diſſolve crude Gold. And
having, for further Tryals ſake, pour'd
ſome of it upon ſpirit of fermented U-
rine, till the Affuſion ceaſ'd to produce any Conflict, and having afterwards gently evaporated away the ſuperfluous moiſture, there did, as I expected, ſhoot, in the remaining Liquor, a
Salt figur'd like Combs and Feathers,
thereby diſcloſing it ſelf to be much of
the nature of Sal Armoniack, ſuch as I
elſewhere relate my having made, by
mingling ſpirit of Urine vvith ſpirit of
common Salt, made the ordinaryway.]
  This (Pyrophilus) is all I can find at
preſent of that Account, of vvhich I
hop'd to have found much more; but
    AA 3   you

you will be the more unconcern'd, for
my not adding divers other things,that,
I remember, I try'd, as vvell before and
after the vvriting the above-tranf-
fcrib'd Paper, (as particularly, that *I*
found the Experiment fometimes to
fucceed not ill, when *I* diftill'd the Oyl
of Vitriol and Sea-falt together, without the intervention of Water, (whereby much time was fav'd,)and alfo when
I imploy'd Oyl of Sulphur, made with
a Glafs Bell, in ftead of Oyl of Vitriol,)
if *I* inform You,that afterwards *I* found,
that *Glauber* himfelf, in fome of his
fubfequent pieces, had deliver'd more
intelligibly the Way of making what
he, without altogether fo great a
Brag,as moft think,calls his *Sal mirabi-*
lis, (which yet fome very ingenious
Readers of his Writings have come to
Us to teach them,) and that thofe Experiments of his about it, which *I* vvas
able to make fucceed, (for fome *I* was
not, and fome *I* did not think fit to try)
you

you will find, together with thofe of
my Own, in more proper places of o-
ther Papers. Onely, to apply what
hath been above related to my prefent
purpofe, *I* muft not here pretermit a
couple of Obfervations.

And firft we may take notice of the
power, that Mixtures, though they feem
but very flight, & confift of the fmalleft
number of ingredients, may, if they make
great changes of Texture, have, in altering the Nature and Qualities of the
compounding Bodies. For in our (above
recited) cafe, though Sea-falt be a Body confiderably fix'd, requires a naked
Fire to be elevated even by the help of
copious additaments of beaten Bricks,
or Clay, &c. to keep it from Fufion,
yet the faline Corpufcles are diftill'd o-
ver in a moderate Fire of Sand, whilft
the Oyl of Vitriol, by whofe intervention they acquire this volatility, though
it be not (like the other) a Groffe or (as
the fame Chymift fpeaks) corporeal falt,

A a 4   but

but a Liquor, that has been already diftill'd, is yet, by the fame operation, fo
fix'd, as to ftay behind, not onely in the
Retort, but, as I have fometimes purpofely try'd , in much confiderabler
heats then That needs in this Experiment be expof'd to. Nor onely is the
oyl of Vitriol made thus far fix'd ,
but it is otherwife alfo no lefs chang·d.
For when the remaining Salt has been
expof'd to a competent heat, that it
may be very drie and white, to be fure
of which, I feveral times do, when the
Diftillation is ended , keep the remaining Maffe (taken out of the Retort and
beaten) in a Crucible among quick
coals, you fhall have a confiderable
quantity (perhaps near as much as the
Sea-falt You firft imploy'd) of a Subftance, which, though not infipid, has
not at all the taft of Sea-falt, or any o-
ther pungent one , and much leffe the
highly corrofive acidity of Oyl of Vitriol.

And

And the mention of this fubftance
leads me to the fecond particular I intended to take notice of, which is a *Phæ-*
nomenon to confirme what I formerly
intimated, That notwithftanding the
regular and exquifite figures of fome
Salts, they may, by the addition of o-
ther Bodies, be brought to conftitute
Chryftals of very differing, and yet of
curious, fhapes. For if You diffolve the
hitherto mention'd *Caput mortuum* of
Sea falt (after You have made it very
dry, and freed it from all pungency of
Taft) in a fufficient quantity of fair water, and, having filtrated the folution,
fuffer the diffolv'd Body leifurely to
coagulate, You will probably obtain, as
I have often done, Chryftals of a far
greater Tranfparency, then the Cubes
wherein Sea falt is wont to fhoot, and of
a fhape far differing from theirs, though
oftentimes no leffe Curious then that
of thofe Cubes; and, which makes
mainely for my prefent purpofe, I have
often

often obferv'd thofe finely figur'd
Chryftals to differ as much in fhape
from one another, as from the Graines
of common Salt. And indeed *I* muft
not,on this occafion,conceal from You,
that whether it be to be imputed to the
peculiar Nature of Sea falt, or (which I
judge much more probable) to the
great difparities to be met with in Liquors, that do all of them pafs for Oyl
of Vitriol, whether (I fay) it be to this,
or to fome other caufe, that the Effect is
to be imputed, I have found my Attempts, to make the beft fort of *Sal mi-*
rabilis, fubject to fo much incertainty,
that though I have divers times fucceeded in them, I have found fo little Uni-
formity in the fuccefs, as made me reckon this Experiment amongft Contin-
gent ones, and almoft weary of medling
with it.

Experiment

Experiment VII. *

I Remember (*Pyrophilus*) I once made an Experiment, which, if I had had the Opportunity to repeat, and had done fo with the like fuccefs, I fhould be tempted to look upon it, though not as a Lucriferous Experiment, (for tis the quite contrary,) yet as fo Luciferous a one, as, how much foever it may ferve to recommend Chymiftry it felf, may no leffe difpleafe Envious Chymifts, who will be troubled, both that one, who admits not their Principles, fhould

---

* Though this VII. Experiment, being confiderable and very pertinent, the Author thought fit to mention it, fuch as it is here delivered, when he writ but to a private friend; yet, after he was induc'd to publifh thefe Papers, twas the (now raging) Plague, which drove him from the Accommodations requifite to his purpofe, that fruftrated the Defigne he had of firft repeating that part of the Experiment, which treats of the Deftruction of Gold: for as for that part, which teaches the Volatilization of it, he had tryed That often enough before.

devife

devife fuch a thing, and that having found it, he fhould not (Chymift like) keep it fecret.

But to give you a plain and naked Account of this matter, that you may be able the better to judge of it, and, if You pleafe, to repeat it, I will freely tell You, That fuppofing all Metals, as well as other Bodies, to be made of one Catholick Matter common to them all, and to differ but in the fhape, fize, motion or reft, and texture of the fmall parts they confift of, from which Affections of Matter, the Qualities, that difference particular Bodies, refult, I could not fee any impoffibility in the Nature of the Thing, that one kind of Metal fhould be tranfmuted into another; (that being in effect no more, then that one Parcel of the Univerfal Matter, wherein all Bodies agree, may have a Texture produc'd in it, like the Texture of fome other Parcel of the Matter common to them both.)

And

. And having firſt ſuppoſ'd this, I fur‑
ther confider'd, That in a certain Men‑
ſtruum,which, according to the vulgar
Chymiſts doctrine, muſt be a worthleſs
Liquor, according to my apprehenſion
there muſt be an extraordinary efficacy
in reference to Gold, not onely to dif‑
folve,and otherwiſe alter it, but to in‑
jure the very Texture of that ſuppoſed‑
ly immutable Metal.

The Menſtruum then I chofe to try
whether *I* could not diſſolve Gold with,
is made by pouring on the rectifi'd oyl
of the Butter of Antimony as much
ſtrong ſpirit of Nitre, as would ſerve to
præcipitate out of it all the *Bezoarticum*
Minerale, and then with a good ſmart
Fire diſtilling off all the Liquor, that
would come over, and (if need be) Co‑
hobating it upon the Antimonial pow‑
der. For though divers Chymiſts, that
make this Liquor, throw it away, upon
Preſumption, that, becauſe of the Ebul‑
lition, that is made by the Affuſion of
the

the fpirit to the Oyl, and the confequent
precipitation of a copious Powder, the
Liquors have mutually deftroy'd or dif-
arm'd each other ; yet my Notions and
Experience of the Nature of fome fuch
Mixtures invites me to prize this, and
give it the name of *Menftruum peracutum*.

Having then provided a fufficient
quantity of this Liquor, (for I have obferv'd that Gold ordinarily requires a
far more copious Solvent then Silver,)
we took a quantity of the beft Gold we
could get , and melted it with 3 or 4
times its weight of Copper, which Metal we choofe rather then that which is
more ufual among the Refiners, Silver,
that there may be the leffe fufpicion,
that there remain'd any Silver with the
Gold, after their feparation; this Mixture we put into good *Aqua fortis*, or
fpirit of Nitre, that all the Copper being diffolv'd , the Gold might be left
pure and finely powder'd at the bottom;
this

this Operation with *Aqua fortis* being accounted the beſt way of refining Gold that is yet known, and not ſubjeɛt, like Lead, to leave any Silver with it, ſince the *Aqua fortis* takes up that Metal. And for greater ſecurity, we gave the Powder to an Ancient Chymiſt, to boile ſome more of the Menſtruum upon it, without communica. ting to him our Deſign. This highly refin'd Gold being, by a competent degree of heat, brought, as is uſual, to its Native Colour and Luſtre, we put to it a large Proportion of the *Menſtruum peracutum*, (to which we have ſometimes found cauſe to adde a little ſpirit of Salt, to promote the Solution,) wherein it diſſolves ſlowly and quietly enough; and there remain'd at the bottom of the Glaſſe a pretty quantity (in ſhew, though not in weight) of white Powder, that the Menſtruum would not touch, and, if I much miſremember not, we found it as indiſſoluble in *Aqua Regis*

Regis too. The Solution of Gold being abstracted, and the Gold again reduc'd into a Body, did, upon a second Solution, yield more of the white Powder, but not (if I remember aright) so much as at the first; now having some little quantity of this Powder, twas easie with Borax or some other convenient Flux, to melt it down into a Metal, which Metal we found to be white like Silver, and yielding to the Hammer, if not to a less pressure, and some of it, being dissolv'd in *Aqua fortis* or spirit of Nitre, did, by the odious Bitterness it produc'd, sufficiently confirm us in our Expectation, to find it true Silver.

I doubt not, but you will demand (*Pyrophilus*) why I did not make other Tryals with this Factitious Metal, to see in how many other Qualities I could verifie it to be Silver, but the quantity I recover'd after Fusion was so small, some of it perhaps being left either in the Flux, or in the Crucible, that I had

not

not wherewithall to make many Tryals,
and being well enough fatisfied by the
vifible Properties, and the Taft peculiar to Silver, both that it was a Metal,
and rather Silver then any other, I was
willing to keep the reft of it for a while,
as a Rarity, before·I made further Tryals with it ; but was fo unfortunate, as
with it to loofe it in a little Silver Box,
where I had fomething of more Value,
and poffibly of more Curiofity.

You will alfo ask, why I repeated not
the Experiment? to which I fhall anfwer,
that, befides that one may eafily enough
faile in making the Menftruum fit for
my purpofe, I did, when I had another
Opportunity, (for I was long without
it,) make a Second Attempt ; and having, according to the above mention'd
Method, brought it fo far, that there
remain'd nothing but the melting of the
White Powder into Silver, when having wafh'd it, I had layd it upon a piece
of white Paper by the fires fide to dry,

being

being fuddenly call'd out of my Cham-
ber, an ignorant Maid, that in the mean
time came to drefs it up, unluckily
fwept this Paper, as a foul one, into the
fire: which Difcouragement, together
with multiplicity of Occafions, have
made me fufpend the Purfuit of this
Experiment, till another Opportunity.
But in the mean time I was confirm'd in
fome part of my Conjecture by thefe
Things.

The firft, by finding, that with fome
other Menftruums which I try'd, and
even with good *Aqua Regis* it felf, I
could obtain from the very beft Gold, I
diffolv'd in them, fome little quantity
of fuch a White Powder, as I was fpeaking of; but in fo very fmall a proportion to the diffolv'd Gold, that I had never enough of it at once, to think it
worth profecuting Tryals with.

The other was this. That a very
Experienc'd Mineralift, whom I had acquainted with part of what I had done,
assur'd

affur'd me, that an eminently Learned and Judicious perfon, that he nam'd to me, had, by diffolving Gold in a certain kind of *Aqua Regis*, and after by redu-ction of it into a Body, rediffolving it again, and repeating this Operation very often, reduc'd a very great, if not much the greater, part of an Ounce of Gold into fuch a White Powder.

And the Third thing, that confirm'd me, was, the Proof given me by fome Tryals that I purpofely made; That the *Menftruum peracutum* I imploy'd, had à notable Operation upon Gold, and would perform fome things (one of which we fhall by and by mention,) which Judicious Men, that play the great Criticks in Chymiftry, do not think feafible: fo that there feems no greater caufe to doubt, that the above mention'd Silver was really obtain'd out of the pure Gold, then onely this, That Men have hitherto fo often in vain attempted to make a real Tranfmutation

B b 2   of

of Metals, (for the better or for the worſe,) and to deſtroy the moſt fix'd and compacted Body of Gold, that the one is look'd upon as an Unpracticable Thing, and the other as an Indeſtructible Metal.

To reflect then a little upon what we have been relating, if we did not miſtake nor impoſe upon our ſelves, (I ſay, upon our Selves, the Project being our own, and purſued without acquainting any body with our Aime,) it may afford us very conſiderable Conſequences of great moment.

And in the Firſt place, it ſeems probably reducible from hence, that however the Chymiſts are wont to talke irrationally enough of what they call *Tinctura Auri*, and *Anima Auri*; yet, in a ſober ſenſe, *ſome ſuch thing* may be admitted, I ſay, *ſome ſuch thing*, becauſe as on the one hand, I would not countenance their wild Fancies about theſe matters, ſome of them being as unintelli-

telligible, as the Peripateticks fubftantial Forms,fo,on the other hand,I would
not readily deny, but that there may be
fome more noble and fubtle Corpufcles, being duely conjoyn'd with the
reft of the Matter, whereof Gold confifts, may qualifie that Matter to look
Yellow, to refift *Aqua fortis*, and to exhibit thofe other peculiar *Phænomena*,
that difcriminate Gold from Silver, and
yet thefe Noble parts may either have
their Texture deftroy'd by a very piercing Menftruum, or by a greater congruity with its Corpufcles, then with
thofe of the remaining part of the Gold,
may ftick more clofer to the former,
and by their means be extricated and
drawn away from the latter. As when
(to explain my meaning by a grofs Example) the Corpufcles of Sulphur and
Mercury do,by a ftrict Coalition,affoci-
ate themfelves into the Body we call
Vermilion, though thefe will rife together in Sublimatory Veffels, without

Bb 3   being

being divorc'd by the fire, and will act,
in many cafes, as one Phyfical Body:
yet tis known enough among Chymifts, That if You exquifitely mix with
it a due proportion of Salt of Tartar,
the parts of the *Alkaly* will affociate
themfelves more ftrictly with thofe of
the Sulphur, then thefe were before affociated with thofe of the Mercury,
whereby You fhall obtain out of the
Cinnabar, which feem'd intenfely red, a
real Mercury, that will look like fluid
Silver.   And this Example prompts
me to mind You, (*Pyrophilus*) That, at
the beginning of this Paragraph, I faid
no more, then that the Confequence, I
have been deducing, might probably
be inferr'd from the Premifes.   For as
tis not abfurd to think, that our Menftruum may have a particular Operation upon fome Noble, and (if I may fo
call them) fome Tinging parts of the
Gold, fo it is not impoffible, but that
the Yellowifhnefs of that rich Metal
may

may proceed not from any particular Corpuscles of that Colour, but from the Texture of the Metal; as in our lately mention'd Example, the Cinnabar was highly Red, though the Mercury, it confisted of, were Silver-coloured, and the Sulphur but a pale Yellow; and confequently, the Whitenefs, and other Changes, produc'd in the new Metal we obtain'd, may be attributed not to the Extraction of any tinging Particles, but to a Change of Texture, whereon the Colour, as well as other Properties of the Gold did depend. But That, which made me unwilling to reject the way, I firft propofed, of explicating this Change of Colour, was, That a Mineralift of great Veracity hath feveral times affur'd me, that a known Perfon in the Relators Country, the Netherlands, got a great deal of Money by the way of Extracting a Blew Tincture out of Copper, fo as to leave the Body White; adding, that he himfelf,

Bb4 having

having procur'd from a friend (to satisfie his Curiosity) a little of the Menstruum, (whose chiefe Ingredients his friend communicated to him, and he to me,) he did, as he was directed, dissolve Copper in common *Aqua fortis*, to reduce it into small parts, and then having kept the *Calx* of the Powder of this Copper for some hours in this Menstruum, he perceiv'd, that the clear Liquor, which was weak in Tast, did not dissolve the Body of the Metal, but onely extract a blew Tincture, leaving behind a very White Powder, which he quickly reduc'd by Fusion into a Metal of the same Colour, which he found as Malleable as before. Which I the lesse wonder at, because the Experienc'd Chymist *Johannes Agricola*, in his Dutch Annotations upon *Poppius*, mentions the making of a White and Malleable Copper in good quantities upon his own knowledge; and that of such a kind of Copper, I have with pleasure made Tryal,

Tryal. I elfewbere relate. But of thefe matters we may poffibly fay more in a convenient place.

The Second thing, that feems deducible from our former Narrative, is, That however moft (for I fay not all) of the Judicionfeft among the Chymifts themfelves, as well as among their Adverfaries, believe Gold too fix'd and permanent a Body to be changeable by Art, infomuch that tis a receiv'd Axiom amongft many Eminent Spagyrifts, that *facilius eft aurum conftruere, quàm deftruere*; yet Gold it felf is not abfolutely indeftructible by Art, fince Gold being acknowledg'd to be an Homogeneous Metal, a part of it was, by our Experiment, really chang'd into a Body, that was either true Silver, or at leaft a new kind of Metal very differing from Gold. And fince tis generally confefs'd, that among all the Bodies we are allow'd to obferve near enough, and to try our skill upon, there is not any, whofe

whose Form is more strictly united to
its Matter then that of Gold; and since
also the Operation, by which the White
Powder was produc'd, was made onely
by a corrofive Liquor, without violence
of Fire, it feems at leaft a very probable
Inference, That there is not any Body
of fo conftant and durable a Nature,
but that, notwithftanding its perfifting
inviolated in the midft of divers fenfible
Difguifes, its Texture, and confequently its Nature may be really deftroy'd,
in cafe this more powerful and appro-
priated Agent be brought by a due manner of Application to work upon the
Body, whofe Texture is to be deftroy'd.

But this Matter we elfewhere handle,
and therefore fhall now proceed to the
Laft and chief Confectaries of our Experiment.

Thirdly then, it feems deducible from
what we have deliver'd, that there may
be a real Tranfmutation of one Metal
into another, even among the perfecteft
and

and noblest Metals, and that effected by
Factitious Agents in a short time, and,
if I may so speak , after a Mechanical
manner. I speak not here of Projection, whereby one part of an Aurifick
Powder is said to turn I know not how
many 100 or 1000 parts of an ignobler
Metal into Silver or Gold , not onely
becaufe , though Projection includes
Tranfmutation , yet Tranfmutation is
not all one with Projection, but far ea-
fier then it: but chiefly becaufe tis not
in this Difcourfe you are to expect what
I can say , and do think , concerning
what Men call the Philofophers Stone.
To reftrain my felf then to the Experiment we are confidering, that feems to
teach us, that, at leaft among inanimate
Bodies, the nobleft and conftanteft fort
of Forms are but peculiar Contrivances
of the Matter, and may,by Agents, that
work but Mechanically, that is, by lo-
cally moving the parts , and changing
their Sizes, Shape, or Texture , be
gene-

generated and deftroy'd; fince we fee,
that in the fame parcel of Metalline Matter, which a little before was true and
pure Gold, by having fome few of its
parts withdrawn, and the reft tranfpof'd,
or otherwife alter'd in their ftru&ure,
(for there appears no token, that the
Menftruum added any thing to the
Matter of the produc'd Silver,) or by
both thefe wayes together, the Form of
Gold, or that peculiar Modification
which made it Yellow, indiffoluble in
Aqua fortis, &c. is abolifh'd, and from the
new Texture of the fame Matter, there
arifes that new Forme, or Convention
of Accidents, from which we call a Metal Silver; and fince Ours was not onely diffoluble in *Aqua fortis*, but exhibited that exceffively bitter Taft, which is
peculiar to Silver, there feems no neceffity to think, that there needs a diftin&
Agent, or a particular A&ion of a Subftantial Form, to produce in a Natural
Body the moft peculiar and difcriminating

ting Properties. For twas but the fame
Menftruum, devoid of Bitternefs, that,
by deftroying the Texture of Gold,
chang'd it into another, upon whofe account it acquir'd at once both Whitenefs in colour, Diffolublenefle in *Aqua
fortis*, and aptnefle to compofe a bitter
Body with it, and I know not how many other new Qualities are attributed.

I know tis obvious to object, that tis
no very thrifty way of Tranfmutation,
inftead of Exalting Silver to the condition of Gold, to degrade Gold to the
condition of Silver. But a Tranfmutation is neverthelefle more or lefle real,
for being or not being Lucriferous, and
fince That may inrich a Brain, that may
impoverifh a Purfe, I muft look upon
your humour as that of an Alchymift,
rather then of a Philofopher, if I durft
not expect that the Inftructivenefs in
fuch an Experiment will fuffice to recommend it to You. And if I could
have fatisfied my felf, that good Authors

thors are not miftaken about what they affirm of the Tranfmutation of Iron into Copper, though, the Charge and Pains confider'd, it be a matter of no Gain, yet I fhould have thought it an Experiment of great Worth, as well as the Tranf-mutation of Silver into Gold. For tis no fmall matter to remove the Bounds, that Nature feems very induftrioufly to have fet to the Alterations of Bodies; efpecially among thofe Durable and al-moft Immortal Kinds, in whofe Con-ftancy to their firft Forms, Nature feems to have defign'd the fhewing her felf in-vincible by Art.

I fhould here(*Pyrophilus*) conclude what I have to fay of the Experiment, that hath already fo long entertain'd us, by recommending to You the repetition of what I had not the Opportunity to try above once from end to end, were it not, that I remember fomething I faid about the *Menftruum peracutum*, may feem to import a Promife of communicating to You

You fomething of the Efficacy of that
Liquor upon Gold. And therefore
partly for that reafon, and partly to
make fure, that the prefent Difcourfe
fhall not be uninftructive to You, I
would adde, That though not onely the
generality of Refiners and Mineralifts,
but divers of the moft Judicious Culti-
vators of Chymiftry it felf, hold Gold
to be fo fix'd a Body, that it can as little
be Volatiliz'd as Deftroy'd, and that
upon This ground, that the proceffes of
fubliming or diftilling Gold to be met
with in divers Chymical Books, are either myftical, or unpracticable, or fal-
lacious, (in which Opinion I think them
not much miftaken;)though Th.s, I fay,
be the perfwafion even of fome critical
Chymifts, yet, upon the juft Expectation I had to find my Menftruum very
operative upon Gold, I attempted and
found a way to Elevate it to a confiderable height, but a far lefs proportion
of Additament, then one that were
not

not fully perfwaded of the poffibility
of Elevating Gold; and though I have
indeed found, by two or three feveral
Liquors, (efpecially the *Aqua pugilum*,
ænigmatically defcrib'd by *Bafilius*,)
that the Fixednefs of Gold is not altogether invincible, yet *I* found the Effect
of thefe much inferior to that of our
Mixture, touching which *I* fhall relate
to You the eafieft and fhorteft, though
not perhaps the very beft, manner of imploying it.

We take then the fineft Gold we can
procure, and having either Granulated
it, or Laminated it, we diffolve it in a
moderate hear, with a fufficient quantity of the *Menftruum peracutum*, and having carefully decanted the Solution
into a conveniently fiz'd Retort, we very gently in a Sand-Furnace diftill off
the *Menftruum*, and if we have a mind
to elevate the more Gold, we either
pour back upon the remaining fubftance
the fame Menftruum, or, which is better,

ter, rediſſolve it with freſh; the Liquor
being abſtraƈted,we urge the remaining
Matter by degrees of Fire, and in no
ſtronger a one, then what may eaſily be
given in a Sand Furnace, a conſiderable
quantity of the Gold will be Elevated
to the upper part of the Retort, and either fall down in a Golden colour'd Liquor into the Receiver, or, which is
more uſual, faſten it ſelf to the Top and
Neck in the form of a Yellow or Red-
diſh Sublimate, and ſometimes we have
had the Neck of the Retort inrich'd
with good ſtore of large thin Chryſtals, not Yellow but Red, and moſt
like Rubies, very glorious to behold;
(though even theſe being taken out,
and ſuffer'd to lie a due time in the o-
pen Air would looſe their ſaline Form,
and run *per Deliquium* into a Liquor.)
Nor ſee I any cauſe to doubt, but that
by the Reaffuſions of freſh Menſtruum
upon the dry *Calx* of Gold, that ſtayes
behind, the whole Body of the Metal

may be eafily enough made to pafs
through the Retort, though, for a certain reafon, I forbore to profecute the
Experiment fo far.

But here (*Pyrophilus*) I think my
felf oblig'd to interpofe a Caution, as
well as to give you a further Information about our prefent Experiment. For
firſt I muſt tell You, that though even
Learned Chymiſts think it a fufficient
proof of a true Tincture, that not onely
the colour of the Concrete will not be
feparated by Diſtillation, but the extracting Liquor will pafs over tincted
into the Receiver; yet this fuppofition,
though it be not unworthy of able men,
may, in fome cafes, deceive them. And
next I muſt tell You, that whereas *I*
fcruple not, in feveral Writings of mine,
to teach, That the Particles of folid
and confiſtent Bodies are not alwaies
unfit to help to make up Fluid ones, *I*
fhall now venture to fay further, That
even a Liquor, made by Diſtillation,

how

how volatile foever fuch Liquors may
be thought, may in part confift of Corpufcles of the moft compact and ponderous Bodies in the World.

Now to manifeft Both thefe things,
and to fhew You withall the Truth of
what I elfewhere teach , *That fome Bo-*
dies are of fo durable a Texture,that their
Minute parts will retain their own Na-
ture, notwithftanding variety of Difgui-
zes, which may impofe, not onely upon o-
ther men, but upon Chymifts themfelves;
I will adde, that to profecute the Experiment, I dropp'd into the Yellow
Liquor afforded me by the Elevated
Gold, a convenient quantity of clean
running Mercury, which was immediately colour'd with a Golden colour'd
Filme, and fhaking it to and fro, till the
Menftruum would guild no more,when
I fuppof'd the Gold to be all præcipitated upon the Mercury, I decanted the
clarifi'd Liquor, and mixing the remaining *Amalgam* (if I may fo call it) of

Gold

Gold and Mercury, with feveral times its Weight of Borax, I did, as I expeced, by melting them in a fmall Cruci. ble, eafily recover the fcatter'd Particles of the Elevated Metal, reduc'd into one little Mafs or Bead of Corporal or Yellow (though perhaps fomewhat palifh) Gold. But yet whether the Gold, that tinged the Menftruum, might not, before the Metal was reduc'd or præcipitated out of it, have been more fuccefully apply'd to fome confiderable purpofes, then a bare Solution of Gold, that hath never been Elevated, may be a Queftion, which I muft not in this place determine, and fome other things that I have try'd about our Elevated Gold, I have elfewhere taken notice of; Onely this further Ufe I fhall here make of this Experiment, that, whereas I fpeak in other Papers, as if there may be a volatile Gold in fome Oars, and other Minerals, where the Mine-men do not find any thing of that Metal, I mention fuch

a

(375)

a thing upon the Account of the paſt Experiment and ſome Analogies. And therefore as *I* would not be underſtood to adopt what every Chymical Writer is pleaſ'd to fancie concerning Volatile Gold; ſo *I* think Judicious men, that are not ſo well acquainted with Chymical Operations, are ſometimes too forward to condemn the Chymiſts Obſervati-ons; not becauſe their Opinions have nothing of Truth, but becauſe they have had the ill Luck not to be warily enough propoſ'd. And to give an in-ſtance in the Opinion, that ſome Mine-rals have a Volatile Gold, (and the like may be ſaid of Silver,) *I* think I may give an Account, rational enough, of my admitting ſuch a thing, by explicating it thus: That *as* in our Experiment, though after the almoſt total abſtracti-on of the Menſtruum, the remaining Body being true Gold, and conſequent-ly, in its own Nature, fix'd, yet it is ſo ſtrictly aſſociated with ſome volatile

C c 3 ſa-

ſaline Particles, that theſe, being preſs 'd
by the fire, carry up along with them
the Corpuſcles of the Gold, which may
be reduc'd into a Maſs by the admiſtion
of Borax, or ſome other Body fitted to
divorce the Corpuſcles of the Metal
from thoſe, that would Elevate them,
and to unite them into Grains, too big
and ponderous to be ſublim'd; *ſo* in
ſome Mineral Bodies there may be pretty ſtore of Corpuſcles of Gold, ſo minute, and ſo blended with the unfix'd
Particles, that they will be carried up
together with them by ſo vehement a
heat, as is wont to be imploy'd to bring
Oars, and even Metalline maſſes to Fuſion.   And yet tis not impoſſible, but
that theſe Corpuſcles of Gold, that in
ordinary Fuſions fly away, may be detain'd and recover'd by ſome ſuch proper *additament*, as may *either* work upon, and (to uſe a Chymical Term) mor-
tifie the other parts of the Maſs, without doing ſo upon the Gold; *or* by aſſociating

.fociating with the Volatile and ignobler
Minerals , fome way or other difable
them to carry away the Gold with
them, as they otherwife may do; *or* by
its Fixednefs and Cognation of Nature
make the difperf'd Gold imbody with
it. On which Occafion I remember,
that a very Ingenious Man, defiring my
Thoughts upon an Experiment, which
he and fome others, that were prefent at
it, look'd upon as very ftrange,namely,
that fome good Gold, having, for a certain Tryal, been cuppell'd with a great
deal of Lead, inftead of being advanc'd
in Colour, as in Goodnefs, was grown
manifeftly paler then before; my Conjecture being, That fo great a Proportion of Lead might contain divers particles of volatile Silver, which,meeting
with the fix'd Body of the Gold, by incorporating therewith, was detain'd,
was much confirm'd by finding, upon
Enquiry, that the Gold, inftead of loofing its Weight, had it confiderably in-

creaf'd

creaf'd; which did much better anfwer my Ghefs, then it did their Expectation, that made the Experiment, and were much furpriz'd at the Event. But this is no fit place to profecute the confideration of the Additaments, that may be uf'd to unite and fix the Particles of the nobler Metals, blended with volatile Bodies; though perhaps what hath been faid may afford fome Hint about the matter, as well as fome Apology for the Chymical Term, Volatile Gold: the poffibility of which, I prefume, we have evinc'd by the latter part of this Experiment, (in which I am forry I cannot remember the proportion of the remaining Salts, that were able to Elevate the Gold;) for That I have feveral times made, and therefore dare much more confidently rely on it, then I can prefs You to do on the former part, (about the Tranfmutation, or at leaft Deftruction of Gold,) till You or I fhall have Opportunity to repeat that Tryal.

Experiment

Experiment VIII.

THough (*Pyrophilus*) the Experiment, I am about to subjoin, may, at the first glance, seem onely to concern the *production of Tafts*, and be indeed one of the principal, that I devis'd concerning that subject, and that belongs to the Notes I have made about those Qualities: yet if You do not of your self take notice of it, I may hereafter have Occasion to shew You, that there are some particulars in this Experiment, that are applicable to more then Tafts. And since I had once thoughts (however since discouraged by the difficulties of the Attempt) to make my *Notes* extend even to *divers Qualities*, which the *operations of Chymists*, and the *practice of Phyficians* have made men take notice of; (such as the powers of *corroding, præcipitating, fixing, purging, bliftering, ftupifying*, &c-) *I* presume You will not dislike, that one, who had
thoughts

thoughts to fay fomething even of
Chymical and of Medical Qualities, if I
may fo call them, fhould give You here
an Experiment or two about more obvious, though *particular*, Affections of
Bodies, when there are feveral things
in the Experiment, that may be of a *general import* to the Doctrine of the O-
rigine of Qualities and Forms.

We took then an Ounce of refined
Silver, and having diffolv'd it in *Aqua
fortis*, wee fuffer'd it to fhoot into Chryftals, which being dried, we found to ex-
ceed the weight of the Silver by feveral
Drachms, which accrued upon the concoagulation of the acid Salts, that had
diffolv'd, and were united to the Metal.
Thefe Chryftals we put into a Retort,
and diftill'd them in Sand, with almoft
as great a heat as we could give in a hammer'd Iron Furnace, wherein the Operation was made; but there came over
onely a very little fowrifh Flegm with
an ill fent, wherefore the fame Retort
being fuffer'd to cool, and then coated,

it was remov'd to another Furnace, capable of giving a far higher degree of Heat, namely, that of a naked fire, and in this Furnace the Diftillation was pursued by the several degrees of heat, till at length the Retort came to be red hot, and kept so for a good while; but though even by this Operation there was very little driven over, yet That fufficiently manifefted what we aimed at, shewing (namely) that a Body extreamly Bitter might afford, as well as it confifted of, good ftore of parts that are not at all bitter, but (which is a very differing taft) eminently Sowr. For our Receiver being taken off even when it was cold, the contain'd spirit smoak'd out like rectify'd *Aqua fortis*, and not onely smelt and tafted like *Aqua fortis*, to the Annoyance of the Nofe and Tongue, but being pour'd upon Filings of crude Copper, it fell immediately to corrode them with violence, making much hiffing, and sending

ding

ding up thick fumes, and in a trice produc'd, with the corroded Copper, a blewish colour, like That, which that Metal is wont to give in good *Aqua fortis*.

Afterwards we took *Minium* and *Aqua fortis*, and made a Solution, which being filtred and evaporated, left us a *Saccharum Saturni*, much like the common made with spirit of Vinegar, then taking this sweet Vitriol of Lead, (as we elsewhere call it) we endeavour'd in the formerly mention'd Sand Furnace to drive it over in a Retort; but finding That degree of fire incompetent to force over any thing save a little flegmatick Liquor, we caus'd the Retort to be coated, and transferr'd to the other Furnace, where being urg'd with a naked Fire, it afforded at length a spirit somewhat more copious then the Silver had done. This Spirit smoak'd in the cold Receiver as the other had, and did, like it, rankly smell of *Aqua fortis*, and was so far from retaining any of the sweetness

nefs of the Concrete that had yielded it,
that it was offenfively acid, and being
pour'd upon *Minium*, it did with noife
and Bubbles fall upon it, and quickly
afforded us a Liquor, which being filtred, did, by its Sweetnefs as well as o-
ther proofs, affure us, that there would
have needed but a gentle Evaporation
(if We had leifure to make it) to obtain from it a true Sugar of Lead; and
tis remarkable, that the Concrete,
which appear'd White before Diftillation, remain'd, for the moft part, behind in
the Retort in the form of a black *Caput*
mortuum, (fometimes We have had it
in a Yellowifh Lump,) which was neither at all fweet, as the Vitriol of Lead
it felf had eminently been, nor at all
fowr, as the Liquor, diftill'd from it,
was in a high degree, but feem'd rather
infipid, and was indeed but a *Calx* of
Lead, which the heat of the fire had in
part reduc'd into true and manifeft Lead
in the Retort it felf, as appear'd by many

ny Grains of feveral Sizes, that We met with in the *Caput mortuum*, (the reft of which is eafily enough reducible by fufion with a convenient flux into malleable Lead it felf.)

There are fome *Phænomena* of this Experiment, that We may elfewhere have Occafion to take notice of, as particularly, That, notwithftanding Silver be a Body fo fix'd in the fire, that it will (as tis generally known) endure the Cuppel it felf, and though in the dry'd Chryftals of Silver, the Salt, that adheres to the Silver, increafes the weight of the Metal but about a 4^h or a 3^d part; yet this fmall proportion of faline Corpufcles was able to carry up fo much of that almoft fixedft of Bodies, that, more then once, We have had the infide of the Retort, to a great height, fo cover'd over with the Metalline Corpufcles, that the Glafs feem'd to be Silver'd over, and could hardly, by long fcraping, be freed from the copious and clofely adhering Sublimate. But

But the *Phænomenon*, that I chiefly desire to take notice of at present, is this, That not onely *Aqua fortis*, being concoagulated with differing Bodies, may produce very differing Concretes, but the same numerical Saline Corpuscles, that, being associated with those of one Metal, had already produc'd a Body eminent in one Taſt, may afterwards, being freed from that Body, compose a Liquor eminent for a very differing Taſt; and after That too, being combin'd with the particles of another Metal, would with them conſtitute a Body of a very eminent Taſt, as oppoſite as any one can be to both the other Taſts; and yet thefe Saline Corpuscles, if, inſtead of this ſecond Metal, they ſhould be aſſociated with ſuch a one as That, they are driven from, would therewith exhibit agen the firſt of the three mention'd Taſts. To prove all this, We took Chryſtals of refined Silver made with *Aqua fortis*, and though thefe Chryſtals be, as We

often

often note, superlatively bitter; yet having, by a naked fire, extorted from them what Spirit we could, and found That, as we expected, extremely Acid, we put one part of it upon a few Filings of Silver, of which it readily made a Solution more bitter then Gall, and the other part of the distill'd Liquor We poured upon *Minium:* and though, whilst it had been an Ingredient of the Chrystals of Silver committed to Distillation, it did with that Metal compose an excessively bitter substance, yet the same Particles, being loosned from that Metal, and associated with those of the Lead, did with them constitute a Solution, which by Evaporation afforded us a *Saccharum Saturni,* or a Vitriol sweet as Sugar. And for further confirmation, We varied the Experiment, having, in a naked Fire, distilled some dry'd *Saccharum Saturni* made with *Aqua fortis,* the little Liquor that came over, in proportion to the Body, that
afforded

afforded it, was fo ftrong a fpirit of Nitre, that for feveral hours the Receiver
was fill'd with red Fumes; and though
the fmoaking Liquor were hugely
fharp, yet part of it, being pour'd upon
a piece of its own *Caput mortuum*, (in
vvhich We perceiv'd not any Taft) did
at length (for it vvrought but very
flowly) exhibit fome little Grains of a
Saccharine Vitriol, but the other part,
being put upon Filings of Silver, fell
upon it immediately vvith noife and
ftore of fmoak, and a vvhile after concoagulated vvith part of it (vvhich it
had diffolv'd) into a Salt exceffively
bitter.

Experiment IX.

THe Artificial Tranfmutation of
Bodies, being as the rareft and dif-
ficulteft Produ&ion, fo one of the nobleft and ufefulleft Effe&s of Humane
skill and power, not onely the clear In-
　　　D d　　　　　ftances

ſtances of it are to be diligently ſought for and priz'd, but even the Probabilities of effecting ſuch an extraordinary Change of Bodies are not to be neglected; eſpecially, if the Verſion, hop'd for, be to be made betwixt Bodies of Primordial Textures, (if I may ſo call them,) and ſuch Bodies, as by the greatneſſe of their Bulk, and by their being to be found in moſt of the mix'd Bodies here below, make a conſiderable part of thoſe, that we Men have the moſt immediately to do with. Invited by theſe conſiderations, *Pyrophilus*, I ſhall venture to give you the *Account* of ſome Obſervations, and Tryals, about the Tranſmuting of Water into Earth, though it be not ſo perfect as I Wiſh, and as I Hope, by Gods bleſſing, to make it.

The firſt Occaſion, afforded me to do any thing about this matter, was my being conſulted by a Gentleman, (an antient Chymiſt, but not at all a Philoſopher,) who relating to me how much he

had

had (with the wonted fuccefs of fuch Attempts) labour'd after the Grand *Arcana*, complain'd to me among other things, that, having Occafion to imploy great quantity of purifi'd Rain-water, he obtain'd from it much *lefs* then he wifh'd of the fubftance that he look'd for, but a *great deal* of a certain whitifh excrementitious Matter, which he knew not what to make of. This gave me the Curiofity firft to defire a fight of it, in cafe he had not thrown it away, (which by good fortune he had not,) and then, taking notice of the unexpe-&ed plenty, and fome of the Qualities. of it, to ask him fome Queftions which were requifite and fufficient to perfwade me, that this Refidenee came not from accidental foulnefs of the Water, nor of the Veffels twas receiv'd in. This I af-terwards often thought of, and indeed it might juftly enough awaken fome fuf-picions, that the little Motes, that have been fometimes obferv'd to appear numerous enough, in pure Rain water whilſt it is diſtilling, might not be meerly accidental, but really produc'd, as well as exhibited by the action of the Fire. I thought it then worth while to proſecute this matter a little farther: And having put a pretty quantity of diſtill'd Rainwater in a clean Glaſs Body, and fitted it with a Head and a Receiver, I ſuffer'd it to ſtand in a Digeſtive Furnace, till, by the gentle heat thereof, the Water was totally abſtracted, and the Veſſel left dry: which being taken out of the Sand, I found the bottom of the Glaſs all cover'd over with a white (but not ſo very white)ſubſtance; which, being ſcrap'd off vvith a Knife, appear'd to be a fine Earth, in vvhich I perceiv'd no manifeſt Taſt, and vvhich, in a vvord, by ſeveral Qualities ſeem'd to be Earth.

This incourag'd me to rediſtill the Rain-water in the ſame Glaſs Body, vvhoſe Bottom, vvhen the Water vvas

all

all drawn off, afforded me more of the
like Earth: but though the Repetition
of the Experiment, and my having, for
greater caution, try'd it all the while in a
new Glafs, that had not been imploy'd
before to other ufes, confirm'd me much
in my conjecture, That unlefs it could
be prov'd, which I think will fcarce be
pretended, that fo infipid a Liquor as
Rain-water fhould, in fo gentle a heat,
diffolve the moft clofe and almoft Indeftructible Body of Glafs it felf, (which
fuch corrofive Menftruums as *Aqua
fortis*, and *Aqua Regis* are wont to leave
unharm'd,) the Earthy powder, I obtain'd from already diftill'd Rain water,
might be a Tranfmutation of fome
parts of the Water into that fubftance,
yet having unhappily loft part of my
Powder, and confum'd almoft all the
reft, (for I kept a little by me, which you
may yet fee,) I fhould, till I had more
frequently reiterated my Experiments,
(which then I had not Opportunity to

D d 3   do,

do, though I had thoughts of doing it
alſo with Snow-water, that I had put
into Chymical Glaſſes for that purpoſe,
and with liquor of melted Hail, which
I had likewiſe provided,) and thereby
alſo obtain'd ſome more of this Virgin
Earth (as divers Chymiſts would call
it) to make farther Tryals with, have
retain'd greater ſuſpicions, if I had not
afterwards accidentally fall'n into diſcourſe of this matter with a learned Phy-
ſician, vvho had dealt much in Rain-
vvater, but he much confirmed me in
my conjecture, by aſſuring me, that he
had frequently found ſuch a White
Earth, as I mention'd, in diſtill'd Rain
Water, after he had diſtill'd the ſame
Numerical Liquor (carefully gather'd
at firſt) I know not how many times
one after another, adding, that he did
not find (any more then I had done) any
cauſe to ſuſpect, that if he had continu'd
to rediſtill the ſame portion of Water,
it would have yielded him more Earth.
But

But the Odnefs of the Experiment
ftill keeping me in fufpence, it was not
without much delight, that afterwards
mentioning it to a very Ingenious Perfon, whom, without his leave, I think
not fit to name, well verf'd in Chymical matters, and whom I fufpeƈted to
have, in order to fome Medicines, long
wrought upon Rain vvater, he readily
gave me fuch an Account of his proceedings, as feem'd to leave little fcruple a-
bout the Tranfmutation we have been
mentioning: for he folemnly affirm'd to
me, that having obferv'd, as I had done,
that Rain-vvater would, even after a
Diftiilation or two, afford a Terreftrial
fubftance, which may fometimes be feen
fwimming up and down in the Limpid
Liquor, he had the Curiofity, being
fettled and at leifure, to try how long
he could obtain this fubftance from the
Water. And accordingly having freed
Rain Water, carefully colleƈted, from
its accidental, and as it vvere fæculent
Earth-

Earthineſs, vvhich it vvill depoſite at
the firſt ſlovv Diſtillation, (and vvhich
is oftentimes colour'd, vvhereby it may
be diſtinguiſh'd from the White Earth
made by Tranſmutatiou,) he rediſtill'd
it in very clean Glaſses, not onely 8 or
10 times, but neer 200, vvithout finding that his Liquor grevv weary of affording him the White Earth, but rather that the Corpuſcles of it did appear far more numerous,or at leaſt more
conſpicuous in the latter Diſtillation,
then in the former. And vvhen I ex-
preſſed my Curioſity to ſee this Earth,
he readily ſhevv'd me a pretty quantity
of it, and preſented me vvith ſome,
vvhich comparing vvith vvhat I had remaining of mine, I found to be exceeding like it,ſave that it vvas more purely
White, as having been, for the main,af-
forded by Rain Water, that had been
more frequently rectify'd. And to
compare this welcome Powder with
That I made my ſelf, I try'd with This
divers

divers things, which I had before try'd
with my own, and (becaufe the quantity prefented me was lefs inconfiderable)
fome others too. For I obferv'd in this
new Powder, as I had done with my
Own, that being put into an excellent
Microfcope, and plac'd where the Sun
beams might fall upon it, it appear'd a
White Meal, or heap of Corpufcles fo
exceeding, not to fay unimaginably,
fmall, that, in two or three choice Mi-
crofcopes, both I and others had occafion to admire it; and their extreme Lit-
tlenefs was much more fenfibly difcern'd, by mingling fome few Grains
of Sand amongft them, which made a
Mixture that look'd like that of Pibble
ftones, and of the fineft Flower. For
our Earth, even in the Microfcope, appear'd to confift of as fmall Particles, as
the fineft Hair-powder to the naked
Eye. Nor could We difcern this Duft
to be tranfparent, though, when the Sun
fhin'd upon it, it appear'd in the Microfcope

ſcope to have ſome Particles a little gli-
ſtering, which yet, appearing but in a
glaring light, we were not ſure to be no
deceptio viſûs. 2. I found, that our
White Powder, being caſt into Water,
would indeed for a while diſcolour it
by ſomewhat Whitening it, which is no
more then Spaud will do, and the fine
duſt of white Marble, and other ſtones,
whoſe Corpuſcles, by reaſon of their
Minuteneſs, ſwimme eaſily for a while
in the Water, but when it was once ſet-
led at the bottom, it continu'd there
undiſſolv'd(for ought I could perceive)
for ſome dayes and nights, as Earth
would have done. 3. Having weigh'd
a quantity of it, and put it into a new
clean Crucible, with another inverted
over it for a Cover, I plac'd it among
quick Coals, and there kept the Crucible red hot for a pretty while; cauſing
the Fire afterward to be acuated with a
blaſt of a Bellows, but taking out the
Powder, I neither found it melted, nor
clotted

clotted into lumps, nor, when I weigh'd
it again, did I fee caufe to conclude that
there was much of it wafted, befides
what ftuck to the fides of the Crucible,
and to a little Clay, vvherewith I had
luted on the Cover, and which (to fhew
you, that the Heat had not been inconfiderable) was in feveral places burnt
red by the vehemence of the fire; and
when *I* afterwards kept this Powder in
an open Crucible among glowing coals,
neither I, nor one that *I* imploy'd to af-
fift me, perceiv'd it all to fmoak; and
having put a little upon a quick Coal,
and blown That too, *I* found that which
I had not blown away, to remain fix'd
(which fome Bodies will not do) upon
quick Coals, that will endure the fire in
a red hot Crucible. 4. *I* found this powder to be much heavier *in fpecie* then
VVater. For imploying a nice pair of
Gold Scales, and a Method that would
be too long here to defcribe, *I* found
that this Powder weigh'd fomevvhat
(though

(though not much) more then twice
fo much common VVater, as vvas equal
to it in Bulk. And leaft fome Corolla-
ries, that feem obvioufly contain'd in
the common, but groundlefs, conceipts
of the Peripateticks, about the Propor-
tions of the Elements in Denfity &c.
fhould make you expect, that this povvder ought to have been much more
ponderous, *I* fhall adde, that having
had the Curiofity, vvhich *I* wonder no
body fhould have before me, to examine
the Gravity of the Earth, which feems
the moft Elementary of any we have, I
took fome fifted Wood-afhes, which I
had cauf'd to be three or four times
boyl'd in a plentiful proportion of Water, to free them from Salt, and having put them very dry into common
Water, I found them but little heavier
then our newly mention'd Powder, fur-
paffing in weight Water of the fame
Bulk but twice, and a little more then a
6th part, (Water and It being very little
more

more then as 1 to 2 $\frac{1}{2}$.) And that you
may the lefs doubt of this, I will yet
fubjoyn, that, examining the Specifick
Gravity of (white) Glafs it felf, I found
that compact Body to be very little, if
at all more then 2 times and a half as
heavy as Water of equal Bignefs to it.
So that the Gravity of that Powder,
which, borrowing a Chymical term, we
have been calling Virgin-Earth, being
added to its Fixtnefs, and other Qualities, it may feem no great impropriety
of Speech to name it Earth, at leaft,
if by Earth we mean not the pure Elementary Earth of the Schools, which
many of themfelves confeffe not to be
found actually feparate, but a Body dry,
cold, ponderous, induring the fire, and,
which is the main, irrefoluble by Water and Fire into other Bodies fpecifically different.

(But to return to the Guife of the
Powder, when I ask'd this Learned man,
whether he obferv'd the Glafs he diftill'd

ſtill'd in to have been fretted by the Liquor, and whether This loſt of its Subſtance, according as it depoſited more
Powder, He anſwer'd me, (and he is a
Perſon of unſuſpected Credit,) that he
found not his Glaſs to have been injur'd
by the Liquor, and that the Water wa-
ſted (though he were carefull it ſhould
not do ſo by Evaporation and Tranſ-
fuſions) by degrees ſo much, that there
remain'd, by his æſtimate, but about
an 8th part of the firſt quantity: and
though, for certain reaſons, he kept by
him the Liquor laſt diſtill'd, yet he
doubted not, but that it might be very
nigh totally brought into Earth, ſince
out of an Ounce of diſtill'd Rain-water
he had already obtain'd near 3 quarters
of an Ounce, if not more, of the often
mention'd Earth.]

Theſe ſeveral Relations will, I ſuppoſe, perſwade You, *Pyrophilus*, that
this Experiment is hopeful enough to
be well worth your purſuing, if not that
per-

perhaps none but fuch a fcrupulous Perfon as I, would think the profecution of
it other then fuperfluous. And if You
do acquiefce in what hath been already
done, you will, I prefume, think it no
mean confirmation of the Corpufcularian Principles, and *Hypo'hefes.* For if,
contrary to the Opinion that is fo much
in requeft among the generality of modern Phyficians and other Learned
Men, that the Elements themfelves are
tranfmuted into one another, and thofe
fimple and Primitive Bodies, which Nature is prefum'd to have intended to be
the ftable and permanent ingredients of
the Bodies fhe compounds here below,
may be artificially deftroy'd, and (without the intervention of a Seminal and
Plaftick power) generated or produc'd:
if, *I* fay, this may be done, and that by
fuch flight means, why may We not
think, that the Changes and Metamor-
phofes, that happen in other Bodies,
which are acknowledg'd by the Moderns

derns to be far more lyable to Alterations, may proceed from the Local Motion of the minute or infenfible parts of
Matter, and the Changes of Texture
that may be confequent thereunto?
Some bold Atomifts would here be de-
·termining, by what particular Wayes
this ftrange Tranfmutation of Water
into Earth may be perform'd, and would
perchance particularly tell you, how the
continually, but flowly, agitated parts
of the Water, by their innumerable occurfions, may by degrees rub, and as it
were grind themfelves into fuch Surfaces, as *either* to ftick very clofe to one
another by immediate contact, (as *I*
elfewhere obferve polifh'd pieces of
Glafs to do,) *or* implicate, and intangle
themfelves together fo, as to make, as it
were, little *knots*; which knots (he would
add,) or the newly mention'd *clufters* of
coherent Particles, being then grown too
great and heavy to be fupported by the
Water, muft fubfide to the bottom in
the

the form of a Powder, which, by reafon
of the fame Gravity of thefe *Molecule*,
and the ftrict Union of the lefler particles that compofe them, obtain an *indifpofition to diffolve in water*, and to
be *elevated or diffipated by the fire*; as
their *Infipidnefs* may be accounted for
by its being but the fame with that of
the Liquor, whence they were made,
and their *Tranfparency* by that of the
Water they were made of, and by the
multitude of the little Surfaces that belong to fo fine a Powder. But though
in favour of fuch conjectures, I could
fomewhat illuftrate them, *partly* by applying to this Occafion what I elfewhere
obferve of the reducing of the fluid Body of Quickfilver by a bare Circulati-
on, (which is but a repeated Diftillation) with a proportionable heat, into a
real Powder, vvhich alfo vvill not fo
eafily be raif'd by the fire, as the fluid
Body, vvhence by change of Texture it
was made; and *partly* by fubjoining, a-

mong other things, how by the conjunction of two diſtill'd Liquors digeſted
together, I have obtain'd good ſtore of
an inſipid Subſtance, that would not diſſolve in Water, and that would long e-
nough indure no inconſiderable degree
of Fire; though, I ſay, by theſe and o-
ther ſuch particulars, I could make our
Atomiſts conjectures leſſe improbable,
yet the full diſquiſition of ſo difficult a
Subject is too long and intricate to be
proper for this place. *

And therefore, without here examining our Atomiſts explication of this *Me-
tamorphoſis*, we will give him leave for
a vvhile to ſuppoſe the Tranſmutation
it ſelf to be real, and thereupon to conſider, whether the Hiſtorical part of it
do not much disfavour ſome of the
chief Doctrines of the Chymiſts, and a

---

*What is here delivered may be, for the main, veri-
fy'd by what the Reader will meet with in the (following) Xth. Experiment, though That be not It which
the Author meant:

fun.

fundamental one of *Helmonts*. For if the
pureſt Water may be turn'd into Earth,
it will not be eaſie to make it improbable, that the other Ingredients of mixt
Bodies, which the Chymiſts call their
Hypoſtatical Principles, are capable of
being tranſmuted into one another,
which would overthrow one of the
main Foundations of their whole Philoſophy; and beſides, if out of the ſim-
pleſt Water it ſelf, a moderate *fire* can
produce a large proportion of Earth,
that was not formally præexiſtent in it,
how ſhall We be ſure, that in all the
Analyſes, which the Fire makes of mixt
Bodies, the Subſtances thereby exhibited are obtain'd by Separation onely,
without any Tranſmutation? As for
Helmont, tis well enough known, that
he makes Water to be the Material
Principle of all Bodies here below,
which He vvould have to be either Water it ſelf, or but Water diſguiſ'd by
thoſe Forms, vvhich the Seeds of

things

things have given it. I will not here examine, whether this Opinion, if he had
reftrain'd it to Animals and Vegetables,
might not, with fome reftriction and ex-
planations, be kept from appearing abfurd, fince my *Eleutherius* hath (though
without abfolutely adopting it) elfevvhere pleaded for its not being fo ex-
travagant, as it hath been thought.

But whereas *Helmont*'s Grand Argument from Experience is grounded on
this, That the Alkaheft doth, as he af-
firms, by being digefted with, and diftill'd from other tangible Bodies, reduce them all at laft into a Liquor, no
way differing from Rain Water, though
we fhould grant the matter of fact, yet
the Experiment of our Powder will
warrant me to queftion their Ratiocina-
tion. For if all mix'd Bodies be therefore concluded to be materially from
Water, becaufe they are, by the Operation of the Fire, and a Menftruum, after
having pafs'd through divers prævious

Changes,

Changes, reduc'd at length into infipid
Water; by the fame way of arguing(and
with greater cogency) I might conclude,
that all thofe Bodies are materially but
difguif'd Earth, fince without intervention of a Seminal Principle, (for *Helmont* will not allow that Title to Fire,
which he ftiles the Artificial Death of
Things) Water it felf may be turn'd into Earth. Indeed if that acute Chymift
were now alive, and had fuch an immortal Liquor, as he defcribes his *Alkaheft*
to be, I would gladly put him upon try-
ing whether that Menftruum would reduce our White Earth into Water. But
there being no more probability of that,
then that fuch reproduc'd Water, being
juft what it vvas before, might be turn'd
into Earth again; it may be probably
faid, that fince thefe Bodies are mutually convertible into one another, (and, as
to the verfion of Water into Earth, by
a feemingly flight Operation,) they
are not either of them ingenerable and

incor-

incorruptible Elements, much lefs the fole matter of all tangible Bodies, but onely two of the Primordial, and of the moft obvious Schematifms of that, which is indeed the univerfal Matter, vvhich, as it comes to have its minute Particles affociated after this or that manner, may, by a change of their Texture and Motion, conftitute, with the fame Corpufcles, fometimes Water, and fometimes Earth.

But (*Pyrophilus*) to leave thefe Reflexions, to return to the bold Conjectures that they are grounded on; though if I had leifure and indulgence enough, I could, I confefs, add many things in favour of fome of thofe Thoughts:* yet I would not have you wonder, that, whilft I vvas mentioning

---

* Of the poffible wayes of turning Liquors into confiftent Bodies, by bending, breaking, twifting, and by otherwife changing the Texture of the Liquor, fee more particularly the *Hiftory of Fluidity and Firmneffe*, publifhd by the Author.

the

the many particulars, that feem to e-
vince the change of Water into Earth,
I fhould let fall fome Words, that intimate a Diffidence about it. For, to dif-
guize nothing unto You, I muft confefs,
that having, in fpight of an unufual care,
unluckily loft a whole paper of the
Powder I had made my felf, and having
unexpectedly been oblig'd to remove
from my Furnaces, before I had made
half the Tryals I judg'd requifite in fo ·
nice a cafe, I have not yet laid afide all ·
my Scruples.

For 1. I would gladly know, whether
the untranfmuted Rain water, by the de-
pofition of fo much Terreftrial Matter,
were grown lighter *in fpecie* then before, or fharp in taft. Next, I would be
throughly fatisfied, (which I confefs I
am not yet, notwithftanding all that the
followers of *Angelus Sala* have confidently enough written,) whether and
hovv far infipid Liquors (as Rain Water is) may, or may not vvork as Men-

E e 4   ftruums

ſtruums upon Stones or Earthy Bodies: not to queſtion, vvhether the Particles of Rain Water may not,by their mutual Attrition, or ſome other action upon one another, be reduc'd into Shapes and Sizes fit to compoſe ſuch a Menſtruum, as the Liquor was not before; as in divers Plants, that ſeem to be nouriſh'd onely with Water, the Sap is endow'd with a ſharp Taſt,and great penetrancy, and activity of parts.

2.It were alſo fit to know,whether the Glaſs Body,wherein all the Diſtillations are made, do looſe of its VVeight any thing neer ſo much, as the obtained Powder amounts to, over and above the Decrement of VVeight, which may be imputed to the action of the Heat upon the ſubſtance of the Glaſs, in caſe it appear by another Glaſs, kept empty in an equal heat, and for the ſame time that the Glaſs looſes by ſuch Operations any thing worth reckoning. And it vvere alſo not impertinent to try, whether

the

the Gravity of the obtain'd Powder be the fame *in fpecie* with that of the Glafs; vvherein the Diftillations were made: (for that it *differ'd but about a* 5th *part* from the weight of Chryftalline Glafs I lately mention'd.) Which Scruple, and fome of the former, I might have pre-vented, if I had had convenient Metal-line Veffels, wherein to make the Di-ftillations inftead of Glafs ones.

3. I could wifh likewife that it were more demonftrably determin'd, what is on all hands taken for granted, (as it appears indeed highly probable,) that diftill'd Rain Water is a perfectly Homogeneous Body, vvhich if it be not, divers fufpicions might be fugge-fted about its Tranfmutation into Earth, and if it be, 'twill be as a very ftrange thing, fo a matter of very great difficulty to conceive, hovv a perfectly and exqui-fitely Homogeneous Matter fhould, without any Addition, or any Seminal and Plaftick Principle, be brought to
af-

ford great store of a Matter of much more Specifick Gravity then it self, since we see, that no Aggregate we can make of Bodies but æquiponderant *in specie* with water, doth, by vertue of their Convention, grow specifically heavier then it.

4. Having had the Curiosity to try, whether Corrosive Liquors would work upon our white Powder, I found, that not onely good Oyl of Vitriol would corrode it, but strong and deflegm'd Spirit of Salt did readily work upon part of it, and that without the assistance of heat, though not without hissing, and exciting great store of bubbles, as I have known such Menstruums do, when put upon *Lapis Stellaris*, or *Ossifragus*, or some such soft Stone; as if that so much defæcated Rain-water, actuated by heat, had resolv'd some of the looser Corpuscles of the Sand or Stone, that, together with some Salts, compose common Glass, as I have observ'd

serv'd in some Petrifying VVater, that
some of the Bodies I took up, and which
were presum'd to be petrify'd, were but
crusted over with 'Stone, that seem'd
generated but by the succeffive appofi-
tion of Stony Particles, that, lying in-
vifibly mingled with the running VVater, stuck in their paffage to the conveniently difpof'd Bodies that lay in the
Streams way. But yet I must not omit,
that, when I fuffer'd this Mixture to fettle, as much of the Powder, as feem'd
to be a very great part of it, remain'd in
the lower part of the Liquor, as if that
had rather fretted then diffolv'd it, and
that not becaufe the Menftruum was o-
vercharg'd or glutted, as I found by putting in afterwards feveral frefh parcels
of Powder, which it readily fell upon,
not without noife and froth. Nor must
I forget, that fometimes I have excited
fuch an Ebullition, by powring the
fame Liquors upon the Earthy part of
Wood-afhes, feveral times wafh'd in

boy-

boyling water, (though, I confefs, I afterwards fomewhat fufpected there
might remain fome little adhering *Al-*
kaly, which might occafion thofe Bubbles, notwithftanding that both I and
another, whom I alfo invited to taft it,
took the Earth to be quite Saltleffe.)
I might (*Pyrophilus*) adde, that fometimes alfo me thought I found this
Powder (which yet likewife fometimes
hapned to me with the lately mention'd
Earth of Wood-afhes) fomewhat gritty
between my Teeth, and fubjoin divers
other particulars, if it were not too tedious to mention to You all the doubts
and confiderations that have occurr'd to
me about the recited Change of Water into Earth: which yet are not fuch as
ought to hinder me from giving You
the Hiftorical account *I* have fet down,
fince to fome of my Scruples *I* could
here give plaufible Anfwers, but that
I cannot do it in few words. And if any
part of our white Powder prove to be
true

true Earth, no body perhaps yet knows
to what the Experiment may lead faga-
cious Men: and whether in a ftrict fenfe
it be true Earth or no, yet the *Phænomena*, that are exhibited in the production of it, are fufficient to give this 9ᵗʰ
Experiment a place among the others
(of the fame Decad) with which tis affociated.   For fince out of a fnbftance
that is univerfally acknowledg'd to be
Elementary and Homogeneous , and
which manifeftly is fluid , tranfparent, much lighter *in fpecie* then Earth,
moift and fugitive, there is artificially
generated or obtain'd a Subftance confiftent, vvhite, and confequently opacous, comparatively ponderous, dry ,
and not at all fugitive; the Alteration is
fo great, and effected in fo fimple a way,
that it cannot but afford us a confiderable Inftance of what the varied Texture
of the minute parts may perform in a
Matter confeffedly fimilar.   And if frequently diftill'd Rain Water fhould not
be

be allow'd Homogeneous, our Experiment will at leaft fhew us, better then
perhaps any hath yet done, how little
we are bound to believe what the Chymifts, and others tell us, when they
pretend manifeftly to exhibit to us Homogeneous Principles, and Elementary
Bodies, and how difficult it is to be certain when a Body is abfolutely irrefoluble into fpecifically differing fubftan-
ces, and confequently what is the detei-
minate number of the perfectly fimple
Ingredients of Bodies: (fuppofing that
fuch there are.) Though I muft confefs,
that my onely aime is not to Relate
what hath been done, but to Procure
the profecution of it. For if the obtain'd
Subftance be, by the Rain Water, diffolv'd out of the Glaffe, this will both
prove a noble and furprizing Inftance
of what may be done by infipid Menftruums, even upon Bodies that are juftly reckon'd among the compacteft and
moft indiffoluble that we know of, and
may

may afford us many other confiderable hints, that have been partly intimated already: and if on the other fide, this Powder, whether it be true Elementary Earth or not, be found to be really produc'd out of the Water it felf, it may prove a *Magnale* in Nature, and of greater confequence then will be prefently forefeen, and may make the Alchymifts hopes of turning other Metals into Gold, appear lefs wild, fince that by Experimentally evincing, that two fuch difficult Qualities to be introduc'd into a Body, as confiderable degrees of Fixity & Weight, (whofe requifitenefle to the making of Gold are two of the Principal things, that have kept me from eafily expecting to find the Attempts of Alchymifts fucceffeful,) may, without the mixture of a Homogeneous Matter, be generated in it, by varying the Texture of its parts.

I will not now adventure to adde any thing of what I have been attempting about

about the tranſmuting (without additaments) of pure Alkalizate Salts into
Earth, becauſe I do not yet know, whether the Tryals will anſwer my Hopes:
(for I do not yet call them my Expe-
ctations.) But upon this ſubject of
Tranſmutations, I could, if it did not
properly belong to another Treatiſe,
tell you ſomething about the Changes,
that may be wrought upon highly rectify'd Spirit of Wine, vvhich vvould
perchance make You think of other
things of the like kind leſſe infeaſible:
For vvhereas tis a known thing, that
That ſpirituous Liquor being kindled,
(and that, if you pleaſe, by other Spirit
of Wine actually fir'd) will, for ought
appears, burn all away, that is, be totally turn'd into flame; if I durſt rely,
in ſo important a caſe; on a couple of
Tryals, whilſt I hope for an Opportunity of making farther ones, I would
tell You, that by a way unthought on
(that I know of) by any Body, I have,
                                    without

vvithout any addition, obtain'd, from
fuch Spirit of Wine, as, being kindled
in a Spoon, would flame all away, without leaving the leaft drop behind it, a
confiderable quantity of downright
incombuftible Flegm. And by another
way (mention'd indeed by *Helmont*, but
not taught to almoft any of his Readers) fome Ingenious Perfons, that you
know and efteem, vvorking by my directions, (but vvithout knowing vvhat
each other vvas doing) did both of them
reduce confiderable quantities of high
rectify'd Spirit of Wine (that vvould
before have burnt all away) into a Liquor, that was for the moft part flegm,
as I vvas inform'd as well by my own
taft, as by the Tryals I order'd to be
made: (being forc'd my felf to be moft
commonly abfent.) From which change
of the greateft part of that at firft liquid
Spirit into Flegm, it feems deducible,
that the fame portion of Matter, vvhich,
by being kindled, may be turn'd all into

Fire,

Fire, may be, by another vvay of hand-
ling, turn'd into Flegm or Water, and
this vvithout the addition of any thing,
and vvithout being vvrought upon by
any viſible Body, but one ſo extremely
dry as duely prepar'd Salt of Tartar; and
that it ſelf is not ſo indiſpenſably neceſſary to the obtaining of flegm out of
totally inflammable Spirit of Wine, but
that, as I was ſaying, I did, by another
way, obtain that dull Liquor vvithout
imploying the Salt, or any other viſible
Body vvhatſoever. But I make a ſcruple to entertain you any longer with
Extravagances of this Nature, and yet,
if I were ſure You vvould contain your
ſmiles, I would adde for concluſion,
That, if I had had time and Opportunity
to furniſh my ſelf with any quantity of
that Water, I had it in my thoughts to
try, vvhether that vvould have afforded
me ſuch a Terreſtrial ſubſtance, as Rain
Water had done, and thereby have un-
dergone a new aud further *Metamor-
phoſis.*   *Experiment*

The X. Experiment.

THere is one Experiment more, two of the chief *Phænomena* of vvhich belong to another Difcourfe; (vvhere I particularly mention Them,) and yet I fhall conclude this little Treatife vvith the recitation of the Experiment it felf, not onely becaufe divers of the *Phæno-mena* do eminently belong to our pre-fent fubjeċt, but becaufe I have fcarce met vvith any Experiments more fui-table to the Defign I have of fhevving, before I conclude this Difcourfe, vvhat great and fudden Produċtions and De-ftruċtions of Qualities may be effeċted by the compofition of the fmalleſt Number of Ingredients, even among Liquors themfelves, and fuch too as are believ'd to be both of Them fimple and Homogeneous, and incapable of Putrefaċtion, that fo it may appear, what notable Alterations of Qualities

even

even seemingly flight and eafie mixtures can perform among Bodies, both
of them fluid, as well as among thofe
that were either both of them ftable, or
one of them ftable, and the other confiftent.

Take then of good Oyl of Vitriol,
and of Spirit of Wine, that will burn
all away, equal parts, not in quantity,
but in Weight; put them together by
little and little, and having plac'd the
Mixture in a Bolt-head, or Glafs Egg
with a long neck, and carefully ftopp'd
it with a Cork and hard Wax, fet the
Veffel in a moderate heat to digeft for a
competent while; (two or three weeks
may do well,) then pour out the Mixture into a tall Glafs Cucurbite, to
which lute on a Head and a Receiver
with extraordinary care, to prevent the
Avolation of the Spirits, which will be
very fubtle: then with a very gentle fire
abftract the fpirit of Wine, that will firft
afcend, and when the Drops begin to
come

come over fowrifh, fhift the Receiver, and continue the Diftillation with great care, that the Matter boyl not over, and when you judge that about half the acid Liquor is come over, it will not be a-mifs; though it be not neceflary, to change the Receiver once more; but whether you do this or no, your Diftil-lation muft be continued, increafing the fire towards the latter end, till you have brought over all you can, and what re-mains in the bottom of the Cucurbite muft be put into a Glafs well ftopp'd, to keep it from the Air.

NB. 1. That to the Production of moft, if not of all the *Phænomena* of this Experiment, it is not abfolutely ne-ceflary, that fo long a Digeftion, (not to fay, not any;)be premif'd; though if the time above prefcrib'd be allow'd, the Experiment will fucceed the better.

2. That, I remember, I have fometimes made ufe of Oyl of Sulphur *per Cam-panam* (as they call it) inftead of Oyl of

Vi-

Vitriol, to produce the recited *Phæno-*
mena; and though the Attempt fucceeded not ill, as to divers particulars, yet
I afterwards chofe rather to imploy oyl
of Vitriol, both becaufe it did, in fome
points, better anfwer my Expectation
then the other Liquor, and becaufe I
would not give occafion to fufpect, that
the Odours, hereafter to be mention'd
as *Phænomena* of our Experiment, were
due to the common Sulphur, whence the
unctuous Liquor, made *per Campanam*,
was obtain'd, as fuch, and did no way
proceed from the acid Vitriolate Salt,
which that Oyl(as tis improperly call'd)
doth abound with.

3.   That I had likewife the Curiofity to digeft Oyl of Vitriol with
Spanifh Wine, inftead of Spirit of
Wine, by which means I obtain'd an
odd Spirit, and refidence, and fome
other *Phænomena*, which I content my
felf to have in this place given hint of,
in regard that Wine being a Liquor of a
much lefs fimple nature then its Spirit,

the *Phænomena*, afforded me by This,
are much fitter for my prefent purpofe.

4. That great care muft be had in re-
gulating the fire, when once a good part
of the Acid fpirit, mention'd in the procefs, is come over. For if the Fire be
not increaf'd, the reft will fcarce afcend,
and if it be increaf'd but a little too
much, the Matter will be more apt, then
one would fufpeɕt, to fwell exceedingly
in the Cucurbite, and perhaps run over
into the Receiver, and fpoil what it finds
there, as it hath more then once hapned
to me, when I was fain to commit the
management of the Fire to others.

Now the oyl of Vitriol, and the fpirit of Wine, being both of them diftill'd
Liquors, and the Latter of them feveral times rediftill'd, and one of them being drawn from fo fimple and familiar a
fubftance as Wine, and the other from
a Concrete not more compounded, then
what Nature her felf (which, as I elfe-
were fhew, can, without the help of Art,
F f 4 produce

produce Vitriol) doth divers times prefent us with; thefe Liquors, I fay, being
both or them diftill'd, and confequently volatile, one would expect, that by
diftilling them, they fhould be brought
over united, as I have tryed, that the
fpirit of Wine, and of Nitre, or alfo of
common Salt may be; and as the fpirits
of differing Vegetables are wont to be;
or that, at leaft, the Diftillation fhould
not much alter them , from what it
found them, after they had been well
mingled together. But this notwithftanding, thefe two Liquors being of
very odd Textures in reference to each
other, their conjunction and diftillation will make them exhibit divers confiderable and perhaps furprizing *Phænomena*.

For Firft, whereas fpirit of Wine has
no great Sent, nor no good one, and
moderately deflegm'd Oyl of Vitriol is
wont to be inodorous; the Spirit, that
firft comes over from our mixture, hath

a Sent not onely very differing from fpirit of Wine, but from all things elfe, that,
I remember, I ever fmelt. And as this
new Odour doth to almoft all thofe,
whofe Opinions I have asked about it,
feem very fragrant and pleafant, fo I
have fometimes had it fo exceeding fubtle, that, in fpight of the care that was
taken to lute the Glaffes exactly together, it would perfume the neighbouring parts of the Laboratory, and would
not afterwards be kept in by a clofe
Cork, cover'd with two or three feveral Bladders, but fmell ftrongly at fome
diftance from the Viol wherein it was
put, I did not think it unlikely, that fo
noble and piercing a Liquor might be
of no mean efficacy in Phyfick; and
though I mifs'd of receiving an account
of its Effects from fome ingenious Phyficians, into whofe Hands I put it to
have Tryals made of it, yet I cannot def-
pair of finding it a confiderable Medicine, when I remember, partly what
hath

hath been done by fome acquaintances of mine with bare flegme of Vitriol, upon the account (as is fuppof'd)of that little Sulphur of Vitriol, that, though but fparingly, doth inrich that Liquor; and partly, what the Mafters of Chymical *Arcana* tell us of the wonderful vertues of the Volatile Sulphur of Vitriol, and what I have obferv'd my felf, that may invite me to have a good Opinion of Remedies of that nature.

2. But to fhevv how much the Odours of Bodies depend upon *their* Texture, I fhall now adde, That after this volatile and odoriferous Spirit is come over, and has been followed by an Acid Spirit, it will ufually, towards the latter end of the Diftillation, be fucceeded by a Liquor, that is not onely not fragrant, but ftinks fo ftrongly of Brim-.ftone, that *I* have fometimes known it almoft take away the Breath (as they fpeak) of thofe, who, when I had the Receiver, newly taken off, in my hand, did

did (either becaufe to make fport I gave them no vvarning, or becaufe they would not take it, as thinking what I told them impoffible,) too boldly adventure their Nofes in the Tryal.

3. There is in this Operation produc'd a Liquor, that will not mingle either with the fragrant, or with the fœtid Spirit hitherto defcrib'd, but is very differing from both of them, and is fo very pleafant, fubtle, and Aromatical, that it is no lefs differing as well from Spirit of Wine, as Oyl of Vitriol. But of this Liquor I give a further Account in a more convenient place.

4. When the Diftillation is carried on far enough, You will find at the bottom, that the two above mention'd Diaphanous Spirits (for Oyl of Vitriol is indeed rather a Saline Spirit, then an Oyl) have produc'd a pretty Quantity of a Subftance, not onely very opacous, but black almoft like Pitch or Jet.

5. And this Subftance, though produc'd

duc'd by two Bodies, that were not onely fluid, but diftill'd, will not alone be
confiftent, but (if the Diftillation have
been urg'd far enough) brittle.

6. And though Spirit of Wine be
reputed the moft inflammable, and Oyl
of Vitriol the moft corrofive Liquor
that is known, yet I could not find, that
this black Subftance would eafily, if at
all, be brought, I fay not to flame, but
to burn; nor that it had any difcernible
Taft, though both the Liquors, from
whofe mixture it was obtain'd, have ex-
ceeeding ftrong and pungent Tafts.

7. And whereas both Oyl of Vitriol and Spirit of Wine will each of them
more readily, then moft Liquors that
are yet known, mingle with common
Water, and diffufe it felf therein, I obferv'd, that this pitchy Mafs, if the Diftillation had been continued till it was
perfectly dry, would not, that I could
perceive, diffolve in common water for
very many hours, and, if I much mifre-
member

member not, for fome dayes:

8. And Laftly, whereas the Oyl of Vitriol, and the Spirit of Wine, were both of them diftill'd Liquors, and one of them exceeding volatile and fugitive; yet the black Mafs, produc'd by them, was fo far fix'd, that I could not make it rife by a confiderably ftrong and lafting fire, that would have raif'd a much more fluggifh Body, then the heavieft of thofe that concurr'd to produce it.

The remaining particulars, that I have obferv'd in this Experiment, belong to another Treatife, and therefore I fhall forbear to mention them in this: nor fhall I at prefent adde any new *Phæno-mena* to thofe I have already recited; thofe frefhly mention'd Experiments, and thofe that preceded it, being, even without the affiftance of the four Obfervations I have delivered before them, fufficient to manifeft the Truth I have been endeavouring to make out. For in the Experiments we are fpeaking of,

it

it cannot well be *pretended*, or at leaſt not well *prov'd*, that any Subſtantial Forms are the Cauſes of the Effeĉts I have recited. For in moſt of the (above mention'd) caſes, beſides that, in the Bodies we imploy'd, the Seminal Ver-tues, if they had any before, may be ſuppoſ'd to have been deſtroy'd by the fire, they were ſuch, as thoſe I argue with would account to be *Faĉtitious* Bodies, artificially produc'd by Chy-mical Operations. And tis not more manifeſt, that, in the produĉtion of theſe Effeĉts, there intervenes a Local Mo-tion, and change of Texture by theſe Operations, then tis inevident and pre-carious, that they are the Effeĉts of ſuch things, as the Schools fancy Subſtanti-al Forms to be: ſince tis, in theſe new Experiments, by the Addition of ſome new particles of Matter, or the Receſs, or Expulſion of ſome præexiſtent ones, or, which is the moſt frequent way, by the Tranſpoſition of Minute parts, yet
without

without quite excluding the other two, that no more skilful a Chymift then I have been able to produce by Art a not inconfiderable number of fuch changes of Qualities; that more notable ones are not ordinarily prefented us by Nature, where fhe is prefumed to work by the help of Subftantial Forms; I fee not, why it may not be thought probable, that the fame Catholick and fertile Prin‑ ciples, *Motion, Bulk, Shape,* and *Tex‑ ture* of the Minute parts of *Matter ,* may, under the Guidance of Nature, (whofe Laws the modern Peripateticks acknowledge to be eftablifh'd by the all-wife God,) fuffice likewife to pro‑ duce thofe other Qualities of Natural Bodies , of which we have not given par‑ ticular Inftances.

F I N I S.

ERRATA.

Praef. p. 11. *l. ult. read* aime. *praef.* p. 13. *l.* 13. *r.* perhaps. *p.* 68. *l.* 13. *r.* deſtroyes. *p.* 130. *l.* 14. *r.* Peare. *p.* 146. *l.* 20. *r.* Principle. *p.* 247. *l.* 25. *r.* Fleurs. *p.* 231. *l.* 15. *r.* it. *p.* 325. *l.* 6. *a Comma at* inflammable. *p.* 337. *l.* 7. *r.* of. *p.* 411. *l.* 7. *r.* former.